PRINCIPLES OF PLANT HEALTH AND QUARANTINE

To my colleagues, past and present, at the former Plant Pathology Laboratory, Harpenden, and at the Central Science Laboratory, Sand Hutton, York, who made my work both rewarding and a pleasure.

David Ebbels

Principles of Plant Health and Quarantine

David L. Ebbels

*Hon. Fellow, Central Science Laboratory
in collaboration with the
Plant Health Group, Central Science Laboratory, York, UK*

CABI Publishing

CABI Publishing is a division of CAB International

CABI Publishing
CAB International
Wallingford
Oxon OX10 8DE
UK

CABI Publishing
44 Brattle Street
4th Floor
Cambridge, MA 02138
USA

Tel: +44 (0)1491 832111
Fax: +44 (0)1491 833508
E-mail: cabi@cabi.org
Website: www.cabi-publishing.org

Tel: +1 617 395 4056
Fax: +1 617 354 6875
E-mail: cabi-nao@cabi.org

© CAB International 2003. All rights reserved. No part of this publication may be reproduced in any form or by any means, electronically, mechanically, by photocopying, recording or otherwise, without the prior permission of the copyright owners.

A catalogue record for this book is available from the British Library, London, UK.

Library of Congress Cataloging-in-Publication Data
Ebbels, D. L.
   Principles of plant health and quarantine / by David L. Ebbels.
     p. cm.
Includes bibliographical references and index.
   ISBN 0-85199-680-9 (alk. paper)
1. Plant quarantine. 2. Plant health. 3. Plant inspection. I. Title.
  SB980.E23 2003
  632'.93--dc21
                      2002155456

ISBN 0 85199 680 9

Typeset in the UK by Wyvern 21 Ltd, Bristol
Printed and bound in the UK by Biddles Ltd, Guildford and King's Lynn

Contents

Preface vii

Acknowledgements ix

1. Introduction to Plant Health and Quarantine 1
2. Early History of Plant Health Control Measures 9
3. International Phytosanitary Controls* 29
4. The European Union Plant Health Regime 48
5. Operation of National Plant Protection Organizations 70
6. Imports and Exports 88
7. Eradication and Containment 114
8. Principles of Certification and Marketing Schemes 143
9. International Certification and Marketing Schemes 171
10. Indexing and Diagnosis in Plant Health 183

*By D.L. Ebbels and A.W. Pemberton.

11. Pest Risk Analysis		212
12. Hygiene and Precautionary Measures		228
Appendix I	The Regional Plant Protection Organizations	252
Appendix II	Convention Concerning the Measures to be Taken Against *Phylloxera vastatrix*, 1878	263
Appendix III	International Controls on the Use of Plant Pests as Offensive Agents	268

Glossary   276

References and Websites   281

Index   293

Preface

In 1997, while leading a Food and Agriculture Organization project in Cyprus, I was involved with organizing and lecturing to a training course for plant health inspectors. In preparing for this course, I realized that the few books available on plant health and quarantine tended to be either collections of more or less specialized papers deriving from scientific meetings, or advanced and detailed texts that were not suitable for the type of course I was giving. This book attempts to fill the need for an introductory text on the principles of plant health and quarantine. The number of people involved with plant health in most countries is relatively small and usually it is not possible to recruit staff experienced in this field. Therefore, whether they are administrators, scientists, or plant health inspectors, there is a continuing need to train new staff while in post, and this book is intended to assist in this process. I also hope that it will be useful for students on university or agricultural college courses in plant protection. The style is intended to be simple, and it is hoped that it will be found easy to follow, even for those whose main language is not English. However, it has been necessary to assume a basic knowledge of biology and to use some specialized agricultural, phytosanitary and scientific terms. A glossary is provided to help in understanding these.

Plant health is a constantly changing field and it would be inappropriate for this book to attempt to cover the regulations of any individual countries, or to give detailed protocols for the latest scientific procedures. Also, it cannot do more than touch on many major fields that have a bearing on plant health, such as biological control and pesticide science. In these cases the references and websites given should provide a lead into the subject where necessary. However, the book does attempt to provide a guide to the international plant health scene and to the basic principles and operations with which plant health officials and scientists will be involved in their day-to-day work. Although it is intended to be applicable worldwide, it is inevitable that in some instances (particularly in Chapter 2) the viewpoint tends to be a European one, and that examples from my own experience will be drawn largely from European agriculture and trade.

David L. Ebbels
Harpenden
November 2002

Acknowledgements

This book would not have been completed without the help of many people, particularly my colleagues in the Plant Health Group of the Central Science Laboratory, a government agency of the UK Department of Environment, Food, and Rural Affairs (Defra), which provides scientific support for the UK plant health administration. There are also others who acted as consultants or who provided information, as noted below. I owe a great debt of gratitude to them all, and thank them sincerely. I am especially grateful to Alan Pemberton who, besides being co-author of Chapter 3, gave me much help and sound advice on most of the other chapters. The people who have acted as consultants, who have provided information, or who have given their expertise in scanning my drafts and eliminating my errors are shown below for each chapter. However, I alone bear the responsibility for any errors that remain.

Chapter 1: Alan Pemberton[1]
Chapter 3: Alan Pemberton[1]
Chapter 4: Paul Bartlett[1], Alison Wright[1], Alan Pemberton[1]
Chapter 5: Stephen Hill[1], Alan Pemberton[1], Stuart Baker[2]
Chapter 6: Alan Pemberton[1], Paul Bartlett[1], Stuart Baker[2]
Chapter 7: Peter Sellar[3], Alan Pemberton[1], Sharon Cheek[1], Joyce Magor[4], Tsedeke Abate[5]
Chapter 8: Peter Reed[1], Aubrey Bould[6], Neil Giltrap[1], John Dickens[3]
Chapter 9: Peter Reed[1], Aubrey Bould[6], Neil Giltrap[1], John Dickens[3]
Chapter 10: Peter Sellar[3], Daphne Wright[1], Sue Hockland[1], Roger Cook[3], Rick Mumford[1], Charles Lane[1], Suzanna Harvey[7]
Chapter 11: Richard Baker[1], Claire Sansford[1], Alan Pemberton[1]

Chapter 12: Alan Pemberton[1], Oliver Macdonald[8]

[1] Central Science Laboratory, York, UK.
[2] Formerly of the Plant Health and Seeds Inspectorate, Defra, UK.
[3] Formerly of the Central Science Laboratory, York, UK.
[4] Formerly of the Natural Resources Institute, University of Greenwich, UK.
[5] Food and Agriculture Organization, Yemen.
[6] Formerly of the Plant Variety Rights and Seeds Division, Defra, Cambridge, UK.
[7] Applied Biosystems, Warrington, UK.
[8] Pesticide Safety Directorate, York, UK.

It is not easy to find suitable illustrations for much plant health work. The images used here come from several sources which are indicated in the figure captions. I am grateful to Mr Don Savage and the Plant Health and Seeds Inspectorate, Defra, UK, for permitting me to use images from their collection. These are Crown copyright and permission to reproduce them should be sought from HMSO Licensing Division, St Clements House, 1–16 Colegate, Norwich NR3 1BQ, UK. I am also grateful to the Central Science Laboratory, York, for permission to use images from the CSL official collection. These are also Crown copyright. I thank Dr T.F. Preece for Fig. 2.1, Catherine Ebbels for Fig. 3.2, Peter Reed for Fig. 8.2, Joe Ostoja-Starzewski for Fig. 10.1, Daphne Wright for Fig. 10.4, Rick Mumford for Fig. 10.7, Richard Baker for Figs 11.1 and 11.2, and the Environment Agency for Fig. 12.1. I also thank Garry Fry and Rob Cook of the Photographic Unit at CSL for scanning and printing the figures.

Introduction to Plant Health and Quarantine   1

Quarantine, Plant Health and Plant Protection

The word *quarantine* derives from the Italian *quarantina*, meaning 'about 40'. After the Black Death arrived in Europe in 1347, observation and experience showed that the incubation time for the disease, from infection to the appearance of symptoms, was a little less than 40 days. This was therefore the period imposed on ships suspected of carrying infection, during which passengers and crew were prohibited from disembarking (MacKenzie, 2001). If any of them showed symptoms, appropriate action could then be taken to prevent disembarkation for a further period, or to refuse entry to the ship and everyone on board. This was a very prudent precautionary and preventive action and, in some circumstances, the same principle is employed today for animals and plants of unknown or suspect health status that arrive at national entry points. However, in some countries the term 'quarantine' has acquired a wider meaning in relation to the prevention of the spread of harmful organisms.

In Europe, the terms 'plant quarantine' and 'plant health' cover much the same subject areas, some countries tending to prefer the former and others the latter. In North America the term 'plant protection' is commonly used. These terms cover the legislative and regulatory measures and associated activities designed to minimize the transport and spread of organisms harmful to plants by means of human activities. There is some variation between countries in just what subject areas these terms are understood to cover. As well as legislative and regulatory measures (sometimes referred to as 'quarantines'), they generally include eradication and containment campaigns, surveys, risk assessment and all closely related topics. Generally the term 'plant health' is also taken to include certification and marketing schemes, and sometimes other less closely connected subject areas, such as the control of migratory pests and the prevention of the misuse of plant pests. This book attempts to cover all these areas, although coverage of the more peripheral topics is relatively brief. A technical term that could also cover these areas is 'phytosanitation'. However, there is also some variation in the interpretation of this term, and for the purposes of this book the term 'plant health' will generally be used.

At its widest, the meaning of 'plant health' also embraces the science of pesticides and their application, registration and regulation. In Europe, the term 'plant protection' covers pesticide science, but can also include the areas of plant quarantine and plant health described above. Pesticides constitute a very wide and well-demarcated specialist field, which this book does not attempt to cover except where it impinges directly on other plant health activities, as with the availability of pesticides for pest eradication campaigns.

Plant Health Terminology

In plant health one is continually dealing with a great variety of organisms potentially harmful to plants, including other plants, fungi, bacteria, viruses, insects, mites, nematodes and members of many other categories of organisms. For simplicity and convenience of reference, therefore, a widely adopted convention is to use the term 'plant pest' to refer to all kinds of organisms harmful to plants, and not only to those belonging to the animal kingdom. This is the sense in which 'pest' is used here, unless another meaning is made clear. Like all scientific disciplines, plant health has a large and distinct terminology of its own and the adjective 'phytosanitary' is usually appropriate to describe matters belonging to this discipline. The phytosanitary terms used in this book are explained where they are first used and are also included in the Glossary. Several international organizations have published more comprehensive glossaries (including those of Hopper, 1995 and Anon., 1996a) but the most widely respected and up to date is the *International Standards for Phytosanitary Measures* No. 5 (Anon., 2002a) the latest edition of which is on the website, www.ippc.int/cds_ippc/IPP/En/default.htm

Weeds and Parasitic Plants

Phytosanitary authorities are usually responsible for control of serious parasitic plant species, such as the dwarf mistletoes (e.g. *Arceuthobium* spp.). However, there is considerable variation between countries in responsibility for control of other plants that may be classified as serious weeds or invasive plants. Although plants that damage crops and other plants by means of competition for space, light and nutrients may fall within definitions of 'plant pest', as being injurious or damaging to plants, they are not always a concern of phytosanitary authorities. Some countries have few or no plant species that are covered by legislative measures for control, while others have many. Responsibility for administering such measures may lie with the phytosanitary authorities or, alternatively, may be the concern of some other part of government, often another part of the agriculture department, the government administration dealing with the environment, or an extension agency.

The Disciplines of Plant Health

As a branch of applied biology, plant health combines much of the disciplines of plant pathology and plant entomology (including nematology and the biology of other invertebrate plant pests). Other branches of biology, and the science of pesticides and their chemistry, also frequently impinge on plant health. Indeed, the endless variety of problems and circumstances that have to be dealt with can involve almost any branch of science, commerce or law, contributing both to its interest and to its intellectual challenge. In short, plant health is the application of scientific knowledge, logic and innovation to administrative and regulatory systems for achieving a good standard of health in plants, including those planted or cultivated and those that constitute the natural vegetation.

Plant health literature

Plant health, as distinct from plant pathology, is seldom given much attention in university undergraduate courses. Also, there are relatively few books and journals devoted primarily to the subject. More recent books include those edited by Hewitt and Chiarappa (1977), Ebbels and King (1979), Kahn (1989) and Ebbels (1993). Many books on plant health, including some of these, consist of papers presented at scientific meetings and often suffer from gaps in subject content and from variable quality in the papers themselves. Other papers on plant health can be found scattered in a wide range of biological and chemical journals. The *EPPO Bulletin*, published by the European and Mediterranean Plant Protection Organization (EPPO), is one of the few specialist journals covering the whole field of plant protection and publishes papers in both English and French. The *FAO Plant Protection Bulletin* was formerly a leading journal in this field, especially in covering legal and administrative matters, but ceased publication after 1994. However, within the United Nations Food and Agriculture Organization (FAO) the Secretariat to the International Plant Protection Convention (IPPC) publishes *International Standards for Phytosanitary Measures* and many other plant health documents, as well as having the informative website already referred to, www.ippc.int/ Other periodicals and documents are published by national government departments (such as the United States Department of Agriculture) and Regional Plant Protection Organizations (RPPOs), particularly EPPO, which are devoted to phytosanitary matters, including pesticides. Many are published in two or more languages, which generally include English. Internet websites are increasingly a major source of basic documents and up-to-date information, and those of the IPPC Secretariat (as above), the World Trade Organization (www.wto.org), the European Commission (www.europa.eu.int), and the RPPOs (see Appendix I) are particularly important in plant health.

The Risk from Alien Pests

Throughout history the movement of human populations has been accompanied by the movement of plants as people carried with them, consciously or unconsciously, the food and forage plants to which they were accustomed and which sustained them and their livestock (Diamond, 1998). Conquests and the establishment of empires greatly facilitated this, and the spread of cultivated plants within the ancient empires of the Middle East, the Egyptians, Greeks and Romans has long been recognized. Along with the plants came many of their associated pests, some occasionally being left behind when they were not transmitted with the seed or other planting material, when soil transmission was avoided, or when they could not establish in the environment of the new area colonized by the human population.

The establishment of trade routes provided a pathway whereby plants and their associated pests could be transported more rapidly. The silk route established between Asia and Europe appears to have facilitated the transport of many crop plants, including apples and oranges from east to west and wheat and barley from west to east. Similarly, trade routes between China, South-East Asia and the Middle East and East Africa facilitated the transport of crops such as rice, sugarcane, soybean and bananas, together with their attendant pests. Although trade had always included plant products, and probably occasionally seeds, by the early 19th century seeds became a frequently traded commodity. In addition to the seeds of food plants, seeds of ornamentals were often gathered by specialist collectors and were sent back to Europe and, later, also to North America, from all over the world. By the middle of the 19th century the long-distance transport of growing plants was becoming more frequent, and this much increased the potential risk that plant pests would be transported together with their hosts. The invention of the Wardian case by Nathaniel Bagshaw Ward in 1830 greatly facilitated the long-distance transport of living plants, enabling them (and some of their pests) successfully to survive the long sea voyages of the time, which were often of many months' duration (Hobhouse, 1992).

Since then, the rapidity of long-distance transport of goods has continually increased, from the introduction of steam ships, which dramatically decreased the length of sea voyages, to modern air freight, by which means living plants can be transported from one continent to another overnight (Fig. 1.1). Containers have facilitated the handling of goods, and the introduction of refrigeration has enabled living plants and any associated pests to be kept fresh and in good condition during transport (Fig. 1.2).

Fig. 1.1. Unloading plant produce from an air freight container. (Photo courtesy of PHSI, Defra. Crown copyright.)

Fig. 1.2. Container transport by land and sea. (a) Delivery of containers to a large container terminal by truck. (b) A specialized container gantry loading containers on to a container ship. (Photo courtesy of PHSI, Defra. Crown copyright.)

The Business of Plant Health and Quarantine

During the past half century or so, these developments have resulted in a huge increase in the movement of plants and plant products, not only through trade but also with war and through the movement of people, particularly with the development of mass tourism during the past few decades. The same developments that facilitated the movement of plants and plant products have also facilitated the transport and spread of many of their associated pests. Recognized pests of all manner of cultivated plants now frequently appear in new areas where previously they were unknown, sometimes with serious consequences for the crops or native vegetation of the area and the economy of the country concerned. Other organisms that are not recognized as damaging in their native areas are also spreading and sometimes become damaging pests in the new areas in which they establish themselves. It is with these threats and risks that plant health and quarantine are largely concerned. Much of the science and work of plant health is devoted to preventing or minimizing the spread and establishment of plant pests in new areas, and to eradicating or controlling them if they do. Laws and regulations, both national and international, aimed at preventing the spread and establishment of plant pests play a major part in this effort. The formulation, administration, enforcement, revision and, where necessary, revocation of such laws and regulations are the basic tasks of those concerned with plant health and quarantine work. This is underpinned by scientific research on the identification and biology of pests and on many other related aspects of their hosts and of pest control.

Established plant pests

Where countries have been trading partners over many decades or centuries, especially if they share similar climates, it is likely that they will have already exchanged any pests of traded plants and plant products, which are able to survive in the partner country. This makes it less likely that such trading partners will now mutually present great plant health risks. However, there may well be serious plant pests already present in a country that are not as yet widespread in all areas where they could survive. In this case, the prevention of further spread, control and eradication of such pests, where possible, are major plant health objectives and these are closely linked with efforts aimed at enabling farmers and growers to obtain healthy and vigorous planting material.

Objectives

It is recognized that different countries approach the objectives and problems described above in different ways, especially with regard to legislation. However, the underlying scientific principles are universal.

The following chapters attempt to explain the science of plant health and quarantine, but without going into great practical detail, as this continually changes and there would be insufficient space here. The current systems used, the many factors that have to be taken into consideration, and the various national and international arrangements that have been developed to prevent plant pests moving with trade are also described.

Early History of Plant Health Control Measures 2

The First Phytosanitary Legislation

The first legislative measures aimed at controlling a plant pest were those concerned with the destruction of the common barberry (*Berberis vulgaris*). Observation and experience in northern Europe had indicated that the 'blast' of wheat and other cereals was noticeably worse in the vicinity of barberry bushes, which were a common constituent of hedgerows (Large, 1940). In many areas this observation engendered a strong belief that the barberry in some way promoted the destruction of the cereals by the 'blast', or black stem rust as it came to be called. This belief in the association of the barberry and the rust was so strong that in 1660 the legislative authorities of Rouen, France, apparently passed a law requiring the destruction of barberry bushes in wheat-growing areas. For the same reason, between 1726 and 1772 legislative measures were also passed by the colonists of New England, requiring or permitting the destruction of barberry bushes in Connecticut, Massachusetts and Rhode Island. Similar measures were taken in Germany in Schaumburg-Lippe (1805) and by Bremen in 1815 (Fulling, 1942, 1943).

It was more than 200 years after the Rouen legislation before the reason for the association between the barberry and the rust of cereals was explained and the belief was shown to be well-founded. In 1865 the renowned mycologist, Anton De Bary, who was at that time Professor of Botany at Freiburg, showed that the barberry acted as an alternate host to the parasitic fungus *Puccinia graminis*, which was the cause of the problem. The barberry was necessary for the fungus to complete its life cycle (Fig. 2.1).

Fig. 2.1. The aecial stage of the black stem rust fungus, *Puccinia graminis*, on the underside of a leaf of common barberry (*Berberis vulgaris*). Aeciospores produced by this life stage of the fungus infect cereals and grasses, resulting in the uredinial and telial life stages and causing the black stem rust disease. (Photo courtesy of Dr T.F. Preece.)

After the scientific basis for action had been established, legislation requiring destruction of the barberry, or enabling local measures for its destruction to be taken, followed in many European states, such as Denmark (1869, 1904), Prussia (1880), Austria (1882), Norway (1916), Bavaria (1920) and Hungary (1920). In most other European countries, especially those in which the black stem rust was not so important, such as in the UK, barberry eradication was left as a voluntary action and was not enforced, although it was sometimes encouraged. In the USA, Congress passed the Plant Quarantine Act in 1912. Led by North Dakota in 1917, this paved the way in 1918 for 13 states to pass their own barberry eradication laws, forming the barberry Eradication Area, which covered virtually all the spring wheat-growing localities in the USA.

Besides *B. vulgaris*, which had been introduced to North America by the early colonists as a hedging plant, many other *Berberis* species, and also several species of the related genus *Mahonia*, were shown to be able to act as alternate hosts of the many subspecies and strains of rusts in the *Puccinia graminis* group. In response to this, in 1919 the United States Department of Agriculture prohibited shipment of 31 species of *Berberis* and four species of *Mahonia* (Fulling, 1942, 1943).

The efficacy of these various measures very much depended on the way the legislation was framed. In some cases the legislation was so hedged about with provisos and exceptions that it was virtually useless, but in others, where the destruction requirements were simple and enforced with few exceptions, it was very effective. This was particularly so in Denmark with the law of 1904, which required complete destruction of barberries except for those in botanical gardens and a few other restricted places. In countries where no legislative action was taken, there seems to be no record of voluntary measures having any noticeable beneficial effect, and it is unlikely that they would have done so unless coordinated on a large scale.

Much later it was found that different genetic forms of the fungus could conjugate on the barberry to produce new strains (Butler and Jones, 1949). It was also later discovered that in warmer climates, such as those of Australia and the southern USA, *Puccinia graminis* could overwinter on winter cereals and living grasses, and so survive from year to year without the aid of the barberry. It is also possible for spores of the fungus to be carried on the wind from such areas to initiate infection in colder climates where overwintering on cereals or grasses is not possible. So control of black stem rust by barberry eradication is not successful everywhere and demonstrates how necessary it is for phytosanitary measures to be soundly based on proven scientific fact. This necessity has only been recognized internationally comparatively recently, but has now been embodied in the principal international agreements on plant health, which insist that regulations that cannot be justified scientifically cannot be maintained (Chapter 3).

The success of the barberry eradication campaign prompted the use of similar legislative measures in the USA to control heteroecious rusts of other economically important species. *Cronartium ribicola* causes the blister rust of five-needle pines, including the white pine, *Pinus strobus*, a major timber tree in eastern North America. Currants and gooseberries (*Ribes* spp.) act as its alternate host. The first Federal Quarantine issued under the 1912 Act in fact sought to prevent the introduction of this disease by prohibiting the importation from Europe of seedlings of four species of five-needle pines. However, prohibition of importation was too late to prevent its introduction. To prevent domestic spread and protect important areas of pines, many states therefore established local Control Areas wherein the destruction of *Ribes* plants was required.

The measures to control the rust of apples (*Gymnosporangium juniperi-virginianae*), for which the alternate host is the common red or 'pencil' cedar, *Juniperus virginiana*, provide another example of such legislative control in the USA. From about 1905 this rust began increasingly to affect certain varieties of apple, and in 1912 was severe in Virginia and adjacent states. In 1914 legislative measures were taken in Virginia to compel the destruction, on request, of red cedar trees within a mile (later increased to 3 miles) of an apple orchard. Six other states followed this lead, with legislative measures to protect their apple orchards against this and related rust diseases (Fulling, 1943).

The Spread of Plant Pests in Trade

The great Swedish natural scientist, Carl Linnaeus, appears to have been the first to express concerns about the risk of spreading plant pests together with their hosts (Usinger, 1964). Linnaeus was aware that a certain beetle was causing considerable damage to pea seeds in North America and that varieties of peas from the English colonies there were being introduced to England. In 1752 he pointed out the possibility that the seed beetles could be introduced to England with the imported pea seed and that from England they might spread and cause damage to peas in continental Europe. He also gave careful attention to scientific methods of controlling many kinds of insects and, in a prize-winning essay (written in 1763 under the pseudonym of C.N. Nelin), he advocated many modern methods, such as fumigation with smokes, oily barriers on the trunks of fruit trees against the winter moth (*Operophtera brumata*), and biological control with a carabid beetle against caterpillars and with *Coccinella* ladybirds, ichneumons and braconid wasps against aphids.

By the middle of the 19th century the causes of disease in plants were becoming clearer. Better land and sea transport were enabling people to travel much more widely and rapidly than had previously been possible, and it also facilitated the exchange of goods, including plants and plant products. In agriculture there was a trend towards larger holdings of land and larger areas of monoculture. Probably, therefore, it was no coincidence that serious plant pests began to appear more frequently in new areas and that their effects became more devastating. The Irish famine of 1845–1848 (Hobhouse, 1992) was the culmination of a series of famines linked to the increasing dependence of the Irish population on the potato (*Solanum tuberosum*) as their only staple food, and to the effects of climate and an introduced plant pest, compounded by irresponsible government. In this case it was due to the appearance of the potato late blight pathogen, *Phytophthora infestans*, which probably reached Europe from its native haunts in Mexico shortly before this, via the newly independent Andean states of South America, in potato tubers carried in the increased volume and speed of trade between the two continents. The trade in guano for use as fertilizer expanded greatly from about 1840, and the ships carrying guano from the Andean states also carried potatoes, so this may well have been the pathway by which the pathogen reached Europe (Bourke, 1964). Nevertheless, devastating as it then was, only one of the pathogen's mating types established in Europe at this time, preventing it from expressing its full genetic potential for variation. Largely due to phytosanitary measures subsequently introduced, it was not until about 1976 that the second, A2, mating type reached Europe and allowed full genetic variation to occur. The consequences of this calamity are now being felt acutely by potato producers in Europe and elsewhere (Spielman *et al.*, 1991; Fry *et al.*, 1993).

In another sector of agriculture on the continent of Europe, vine cultivation had become big business by the middle of the 19th century and the area under vines in Europe was huge. Trade in vine planting material between Europe and other vine-growing areas, especially North America, had increased substantially, so again it was not surprising that another North American pest, this time of vines, made its appearance in Europe in about the year 1865.

Development of International Phytosanitary Agreements

Significantly, it was not famine but the threat to the wine industry that gave rise to the first international measures against a plant pest. Between 1865 and 1875 a huge disaster struck the French wine industry. This was the establishment and spread of the American vine louse (*Viteus vitifolii*) in French vineyards. Still often better known by its former name, *Phylloxera vastatrix*, this pest rapidly spread throughout the vine-growing districts of not only France but the whole of Europe and then most of the rest of the world. Losses in France alone were assessed as the equivalent of £50 million sterling in 1875 (Large, 1940). It is difficult to overemphasize the magnitude of this problem to the wine industry of that time, an industry that was not only of great commercial importance but also dear to the hearts of most vine-growing Europeans. It was recognized that the pest had been carried from North America to Europe on vine material intended for use in hybridization, and that its further long-distance spread had been due to the dissemination of infested planting material by humans. This was the impetus that first brought interested countries together in Berne, Switzerland, to discuss what might be done. Representatives of Austria, France, Germany, Italy, Portugal, Spain and Switzerland attended and the outcome, the *International Convention on Measures to be Taken against* Phylloxera vastatrix, signed on 17 September, 1878, became the first international agreement designed to prevent the spread of a plant pest (Anon., 1914).

The 1878 Convention embodied many of the principles that are recognized today in international plant health. The most important of these were:

1. The responsibility to give an official written assurance on the *Phylloxera*-free (i.e. pest-free) provenance of host material being traded internationally;
2. The prohibition of international trade in certain kinds of material that might spread the pest;
3. The designation of official bureaux responsible for administering such trade;
4. Powers to inspect traded material and to take remedial action on items not complying with the requirements of the Convention;
5. The prompt exchange of relevant information, particularly on new outbreaks; and
6. That all these measures were to be embodied in national law by the participating countries.

A translation of the text is given in Appendix II.

Use of the Convention during the next 3 years highlighted various deficiencies, particularly in relation to lack of clear definitions of terms. This prompted reconvening the international meeting, which reached agreement on a second Convention in 1881. This second Convention, also signed in Berne, was in effect a revision and extension of the earlier one, and it also contained definitions of terms that had evidently caused problems in interpretation. Eight years later, in 1889, a third Convention was signed at Berne.

Much progress was made in the sciences of plant pathology and entomology in the later years of the 19th century and the early years of the 20th century. Plant pests and their effects on crops were more acutely observed and widely recognized. In addition, the continued increase in international and, in particular, intercontinental trade in plants and plant products also led to the spread of various pests. By this time more governments were beginning to take an interest in preventing the spread of serious plant pests, and national legislation was beginning to appear.

Colorado Beetle

The Colorado beetle (*Leptinotarsa decemlineata*, Fig. 2.2) has been responsible for many developments in plant health over the past 150 years. Its high potential for destruction of the potato (*Solanum*

Fig. 2.2. The Colorado beetle, *Leptinotarsa decemlineata*, a serious pest of potatoes (*Solanum tuberosum*), was first described by Thomas Say from the area of the 'upper Missouri' in northern Colorado, USA, in 1824. (Photo courtesy of CSL. Crown copyright.)

tuberosum), its main crop host, and its capacity for long-distance spread along trade routes, on vehicles as well as with plants and plant produce, combine to give it a very high profile as a plant pest. Colorado beetle was first described from the Colorado area in 1824 by Thomas Say, naturalist to Long's expedition to the Rocky Mountains. By the middle of the 19th century it had spread widely along the wagon routes and had become a recognized pest of potatoes in the USA. Charles V. Riley, the Missouri State Entomologist, described its rapid spread eastwards to reach the eastern seaboard in 1874, and forecast that 'even the broad Atlantic may not stay its course' (Riley, 1877). This proved prophetic sooner than even he could have anticipated as in 1876 it was found breeding in potato fields near Bremen. Subsequently it was found elsewhere on the continent of Europe, especially near major ports but also, notably, near Torgau in Germany, south of Berlin. However, these colonies and non-breeding individuals were eliminated successfully (Bartlett, 1979).

The finding of the outbreak near Bremen was evidently a major alert for the British agricultural authorities. Realization of the damage that Colorado beetle might do if it successfully established itself in Britain, especially when memories of the Irish famine were still alive, evidently prompted rapid action. Phytosanitary legislation in the UK started with The Destructive Insects Act, 1877. In fact, despite its general title, this legislation was concerned only with the Colorado beetle, and was a measure of the threat that the British government felt this pest represented. However, the first British outbreak did not occur until 1901 (in allotment gardens at Tilbury Docks near London) and this was eradicated successfully.

Phytosanitary Legislation in the UK

Yet another emigrant from North America reached central Europe in 1890, being carried on gooseberry plants (*Ribes uva-crispa*) imported to south-western Russia. This was the American gooseberry mildew (*Sphaerotheca mors-uvae*), which was much more aggressive than the native European kind (*Microsphaera grossulariae*) and spread as an epidemic (epiphytotic) throughout Europe. Without any means of mildew control, severe economic damage resulted, some crops being completely unsaleable. The gooseberry crop in Britain was more important then than now, and when the disease reached Ireland in 1900 widespread concern prompted considerable efforts to prevent its entry to Great Britain on imported plants (Large, 1940). One result was that in 1907 the Destructive Insects Act, 1877 was replaced by the Destructive Insects and Pests Act. The 1877 Act, although targeted specifically at Colorado beetle, had in fact established many of the basic principles and methods of administrative pest control. It authorized the prohibition or regulation of landing of potatoes or other vegetables, the destruction of an infested crop or one to which the pest might spread, conferred powers for control work on local authorities, and prohibited the keeping of live pests (beetles). The 1907 Act extended these provisions and widened coverage to include the American gooseberry mildew and other pathogens. It also introduced the concept of notification, whereby anyone who suspected the presence of a pest scheduled under the Act was legally bound to report it to the plant health authorities. All these principles are now accepted in phytosanitary legislation and practice in many countries.

One of the measures taken under the 1907 Act was to set up in England a system of county inspectors for American gooseberry mildew, to try to detect and eradicate outbreaks at an early stage. This body of inspectors was in fact a precursor, even if not the direct antecedent, of what is now the Plant Health and Seeds Inspectorate, the operational arm of the national plant protection organization in England and Wales. Nevertheless, the American gooseberry mildew campaign was unsuccessful, and within a few years of the first recorded English outbreak in 1906 it was present in all areas growing gooseberries, or black- or redcurrants. What the British agricultural authorities had overlooked were the enormous powers of spread possessed by all mildews and many other groups of fungi with air-borne spores. No matter how efficiently or how soon the inspectors detected and eliminated new outbreaks, they were always one step behind the mildew, which, by the time it was detected, had already spread by its wind-borne conidia to start new infections (Salmon, 1914). Once such a pest has achieved a foothold in a new host area, it is not amenable to control by administrative means. This lesson was applied more recently in the UK in the case of cucumber downy mildew (*Pseudoperonospora cubensis*). Despite modern effective fungicides and the fact that the disease was confined to glasshouses and apparently did not overwinter, the campaign against this disease was discontinued in 1991, when it was realized that new undetected infection was arriving not only on imported young cucumber plants, but also very possibly as wind-borne sporangia from outbreaks on continental Europe (Ebbels, 1990).

Economic Damage

Like its major pest, the Colorado beetle, the potato occupies a key place in the development of plant health and, indeed, of the science of plant pathology. The fact that it is one of the most important crops in many agricultural systems, that it is vegetatively propagated, that it is susceptible to a large number of pests from all the main groups of organisms harmful to plants, and that it was spread by human from South America to other continents, leaving some of its associated pests behind, all contribute to making the potato crop the subject of many advances and innovations of plant health science and technology, as well as some of its greatest disasters.

Like the Colorado beetle, *Synchytrium endobioticum* is a pest of the potato. It is a primitive fungus belonging to the Order *Chytridiales* and it attacks the young tubers just as they start to form at the ends of the stolons, causing a cauliflower-like proliferation of tissue instead of a normal tuber (Fig. 2.3). Stems and other parts of the plant may also be attacked and tubers are often damaged by being partially affected. This disease came to the notice of potato cultivators in the wetter parts of Europe (and probably also on Prince Edward Island, Canada) during the 1870s. However, it was not described scientifically until 1896 (from

Fig. 2.3. Potato tuber (*Solanum tuberosum*) with wart disease caused by the fungus *Synchytrium endobioticum*. (Photo courtesy PHSI, Defra. Crown copyright.)

Hungary), and it was many years before the life history of the fungus was fully worked out. In Britain the disease was known as the tumour or black scab, and later as the wart disease. It appeared to spread rapidly after it was first recognized in Scotland in 1901 and in north-west England in 1902, probably carried in the developing trade in seed potatoes from Scotland and the north of England to more southerly areas of England and Wales. Although there were many outbreaks, most were in allotments and gardens, and comparatively few were in commercial crops. Nevertheless, the general concern in the British agricultural community and farming press was noticed by other countries, which began to take stringent measures to limit or prohibit the importation of British potatoes, causing much greater economic losses than those caused directly by the disease itself (Large, 1940). This prompted action under the Destructive Insects and Pests Act, 1907, and the subsequent wart disease Order of 1910. The English Board of Agriculture appointed inspectors, who included in their duties the investigation of wart disease outbreaks and the enforcement of the control measures required under the legislation.

Certification Schemes

Almost as soon as wart disease had been recognized in Great Britain, it was discovered that several varieties of potato were apparently immune to infection. Varieties such as Abundance, Snowdrop, Golden Wonder, Langworthy and Conquest enabled potato cultivation to continue in areas where the disease was severe. However, there was a problem: the nomenclature of potato varieties was in complete chaos. The same variety could be known by many different local names in different districts, and the same name was sometimes used in different places for different varieties. Furthermore, stocks often contained a mixture of different varieties, so that sometimes it was doubtful as to which variety the stock was supposed to be. This made it very difficult for those who wished to obtain planting stock of varieties immune to wart disease, who might find that their purchased stock was either not the variety desired or, worse, contained admixture of a variety susceptible to wart disease. The same problem also applied to other characteristics. For example, some potato varieties matured early and others late, while some were good for crisping and others were not. If one could not rely on the purchased stock being almost uniformly the variety it was supposed to be, serious economic losses could result. It was also recognized that seed potatoes were often unhealthy and it was felt that there could be some carry-over of disease from the tubers planted to the crop grown from them.

This problem was not a peculiarly British one but had also been recognized elsewhere, and in Germany a solution had been found which was well established before 1911 under the direction of Dr Otto Appel. This idea was to inspect seed potato crops during the growing season for signs of disease (particularly the then poorly understood leaf rolling diseases). If less than 5% of plants showed symptoms, an official certificate of approval was awarded to the grower, who could then use it to sell the seed potatoes at an enhanced price. This also benefited the purchaser, who thus had some independent assurance that the potato stock he was buying was reasonably healthy. Applications for the inspection of seed potato crops were made to the Deutsche Landwirtschafts-Gesellschaft or to the local Landwirtschaftkammer, and the cost of the inspections was borne by the applicant. This certification system was noted and enthusiastically promoted in the USA and Canada by W.A. Orton, H.T. Güssow and Paul A. Murphy, among many others. Certification systems for seed potatoes were established in these countries between 1913 and 1915 in the seed potato-growing areas, such as Maine, Wisconsin, Prince Edward Island and Nova Scotia (Rieman, 1956). In fact, improvements on the German system were made, in that inspections and certification criteria took into consideration the vigour of the crop and its trueness to variety, as well as the incidence of disease. In Canada, during the decade of the 1920s, the average yield of potatoes from certified seed was approximately double that from uncertified seed, and this is indicative of the success of these schemes (Ebbels, 1979; Ainsworth, 1981).

In The Netherlands there is a long history of progress in the development of quality-control processes for planting material. The Nederlandse Pomologische Vereniging (Dutch Pomological Association) was established in 1898 with the aim, amongst other things, of providing some assurance of quality and trueness to variety for plants marketed by its members. This was not very successful, but in the following years agricultural associations in many parts of The Netherlands introduced quality inspections for planting material, each with their own rules. In 1919 two national organizations were formed, the Centraal Comité voor de Keuring van Gewassen in Nederland (Central Committee for the Inspection of Plants in The Netherlands), and the Keuringsinstituut voor Zaaizaad en Pootgoed (Inspection Institute for Seed and Planting Materials). Both organized and conducted inspections of seed, which caused undesirable competition and confusion. To avoid this, the Nederlandse Algemene Keuringsdienst voor Zaaizaad en Pootgoed van Landbouwgewassen (NAK; Netherlands General Inspection Service) was formed in 1932 to provide quality inspections of seed potatoes and true seed of agricultural field crops (Anon., 2000).

In England the question of potato variety names was tackled in 1915. Trials of potato varieties for susceptibility to wart disease were repeated on a more extensive scale than had been done previously in 1909. In the course of this work variety characteristics and any apparent synonymy were also noted. In 1916 the work was taken over by the group that became the Potato Synonym Committee of the newly established National Institute of Agricultural Botany, and over the next few years the nomenclature of potato varieties was largely clarified (Ebbels, 1979).

This cleared the way for the certification of seed potatoes in Great Britain, which was introduced in Scotland in 1918 with the main aim of authenticating wart disease-immune varieties as true to type. Disease and vigour were not considered at first. The scheme was extended to England in 1919, and was so successful in authenticating stocks as true to variety that non-immune varieties were included from 1922 in Scotland and from 1924 in England. The scheme was also developed in 1922 to comprise two grades of quality: the lower had a 3% tolerance for rogues (plants not of the nominal variety); while the Stock Seed grade was reserved for stock of 'exceptional health and purity'. The idea of having different grades of certification was adopted in many countries. During the next decade, new grades of increasingly high purity were introduced from time to time in Great Britain and eventually, in 1932, the assessment of the crop for virus diseases was incorporated in the certification system in Scotland. However, this did not become a part of the English scheme until 1940 (Ebbels, 1979).

Testing and certification of field crop seeds

Testing of seeds for purity (to ascertain whether they would germinate reasonably well and whether there was admixture of seeds of other varieties or species, including weeds) was started first in Denmark and Germany in 1869. It may well be that in Germany this gave rise to the development of certification of seed potatoes, as described above. Regional seed testing congresses were held at Hamburg in 1906, and at Munster and Wageningen in 1910. After the First World War a larger congress, comprising 16 countries, was held in Copenhagen in 1921, followed in 1924 by a congress of 28 countries at Cambridge, England. Agreement at this congress established the International Seed Testing Association (ISTA), with the aims of promoting standard procedures for sampling and testing seeds and their uniform adoption for seeds moving in international trade. At a very early stage, ISTA established a Plant Disease Committee to investigate and promote methods for the testing of seeds for health although, as explained in Chapter 8, for many crops health has not been a prime objective of seed testing and certification. However, regular workshops for seed pathologists have been held under the auspices of the ISTA Plant Disease Committee since 1958. Under its impetus, official seed testing stations were established in countries and territories that were members of the ISTA, which were responsible for testing seeds and issuing official certificates of quality. Seed testing and certification has been practised and promoted particularly by Denmark, which has remained in the forefront of developments in this field since its inception (Thompson, 1974; Ainsworth, 1981).

International Congresses

In the last quarter of the 19th century, the prosperity of agriculture declined worldwide, and by the end of the century agriculture was in a very depressed state. At the end of the 19th century and the beginning of the 20th, international concern about this situation was reflected in congresses dealing with agricultural matters, which were held at intervals of a year or so in various capitals. At the International Congress of Agriculture and Forestry held in Vienna in 1890 a proposal was made to form an International Phytopathological Committee, partly to coordinate efforts to minimize and control the evident spread of pests in trade. This was followed up at the International Congress at The Hague in 1891 by Professor Rostrup, who drew attention to the need for a system for preventing the introduction of pests on living plants or seeds. However, an International Phytopathological Committee was not established until the International Agricultural Congress of 1903 in Rome. A major concern at this time was the San José scale insect

(*Quadraspidiotus perniciosus*), and Dutch tree-growers in particular were concerned that their profitable export trade to the USA would be threatened by this pest. This led to the establishment of the Plantenziektenkundige Dienst (The Netherlands Plant Protection Service) in 1899, with the aim of gathering and supplying information to government and growers, and of formulating and enforcing phytosanitary regulations. Ritzema Bos, professor of phytopathology at the University of Amsterdam, became the first general manager (Koeman and Zadoks, 1999). With a general increase in awareness of the need to prevent the spread of serious plant pests in international trade, many countries introduced legislation and regulations with this aim during the first two decades of the 20th century.

The International Institute of Agriculture

In 1885 the market price of fruit in California was below the cost of picking and packing, which reflected the collapse of agricultural prices worldwide. This severely affected David Lubin, an American citizen of Polish origin, who had just started a farming and fruit-growing enterprise in California. David Lubin was then aged 36 and had become wealthy through starting a department store and mail-order business in Sacramento in 1874, after trying various other less rewarding occupations. He believed that it was not overproduction that had ruined the farmers' markets but lack of information that would have enabled them to market their produce in a favourable manner. He organized the California Fruit Growers' Exchange and recommended sale by auction, as at Covent Garden in London. Lubin became very active in the USA in forming schemes and lobbying for measures aimed at improving the economics of farming and in 1896 was invited to address the International Agriculture Congress in Budapest on his ideas for an International Agriculture Organization.

By 1904 David Lubin was even more convinced of the necessity for an organization to gather and issue information and coordinate agricultural interests on a global basis, and he travelled to Europe to lobby European governments. The idea, being both radical and expensive, and coming from someone not well known in Europe, was not well received. Eventually he decided to put the proposal before King Victor Emmanuel III of Italy. Overcoming strong discouragement from the Italian government, he was eventually granted an audience by the King, who was impressed with the ideas and the fervour and conviction of this charismatic American and agreed to sponsor the project. King Victor Emmanuel III put the proposal to the Italian Prime Minister on 24 January, 1905. An international conference in Rome in May 1905 resulted in support from 40 nations, who signed a convention giving birth to the establishment in Rome of the International Institute of Agriculture (IIA). It was several years before adequate funding was agreed and the Secretariat, headed by a Secretary-General, did not start work until 1908, covering the areas of agricultural information, statistics, economic and social studies, a legal service and a library (Anon., 1969). The institute had offices in the Villa Borghese area until, in 1932, accommodation for the IIA was constructed by the Italian government on the street now named Via Lubin. This building has since reverted to Italian government use.

International conferences on plant pathology and on plant protection hosted by the IIA in 1914 and 1929 resulted in the International Convention for the Protection of Plants, 1929 (Anon., 1914, 1929; Rogers, 1914) However, this Convention was overtaken by the deteriorating international situation and subsequently the Second World War. It was ratified by only 12 of the 46 signatory countries, although another five adopted its principles without ratification (Chock, 1979). The non-political nature of the IIA preserved it through the First and Second World Wars, during which representatives of belligerent nations sometimes attended sessions side by side.

Control of Potato Wart Disease in the UK

The economic effects of bans by several countries on imports of seed potatoes from the UK for fear of wart disease encouraged efforts to control its occurrence and spread. This was done from opposite ends of the production chain, trying both to control spread of the disease in the field and also, by means of certification, to impose quality control on the seed potatoes sold, although at this time certification of seed potatoes remained voluntary. Where there were large numbers of outbreaks, The Wart Disease of Potatoes Order, 1914, established Infected Areas in which only immune varieties were licensed to be planted and, later, The Seed Potatoes Order, 1918, took control of seed potato sales. The Seeds Act, 1920, updated these provisions and, in effect, made certification of seed potatoes almost compulsory. The Wart Disease of Potatoes Order, 1923, made the disease notifiable, prohibited the sale of affected tubers, permitted the planting only of immune varieties on affected land and (with certain exceptions) banned the movement of potatoes from Infected Areas to areas not so declared. However, licensing policy was relaxed to permit the planting of susceptible varieties on clean land in Infected Areas. Commercial crops were still only rarely infected, the main damage being to trade, and the vast majority of the very large number of wart disease outbreaks remained in private gardens. However, the measures taken proved very effective and the rate of spread of the disease declined dramatically (Pratt, 1979).

Further Development of Crop Certification Schemes

Meanwhile, the success of the certification system of quality control in seed potatoes soon prompted its application to other vegetatively propagated crops. Similar difficulties of variety identification and susceptibility to systemic diseases made strawberry plants and blackcurrant bushes suitable candidates for the first fruit plant certification schemes, which were started in England and Wales in 1927. In Scotland, schemes for the certification of blackcurrant plants and raspberry canes were in operation by 1930. Again, the main objective was the authentication of variety and purity of the stock (in freedom from admixtures). Health was not at first an important consideration, partly because knowledge of the virus diseases of these crops was at best rudimentary but, as plant virology progressed, stocks that were unthrifty or with prevalent symptoms of infection were not certified (Ebbels, 1979). In The Netherlands, flower bulb growers' associations had also instituted quality inspections for the principal bulb crops, and these were later amalgamated into the Bloembollenkeuringsdienst (the Dutch Flower Bulb Inspection Service).

In Great Britain, the advance in knowledge of plant disease organisms was also starting to make the Destructive Insects and Pests Act, 1907, out of date and a revised Act of the same title was passed in 1927 that extended the existing powers to cover bacteria and other disease-causing organisms. This enabled statutory action to be taken against viruses also, and made some new financial provisions concerning compensation.

Colorado Beetle in Europe

The Colorado beetle situation in Europe changed significantly when an extensive infestation was discovered in 1921 near Bordeaux. This probably became established during the First World War, when attention was focused on other matters, and there is some evidence that it may have been transported to the area along with the American forces that were stationed there. Measures to prevent the spread of Colorado beetle in continental Europe were vigorously (but unsuccessfully) pursued and British concerns were reawakened. New Colorado Beetle Orders in 1923 and 1924 required all imported living plants, including potato tubers and tomato fruits, to have been produced more than 40 km from an infested area. The spread of the French outbreak necessitated successive amendments to this legislation. Environmental conditions in the mid 1930s were evidently conducive to Colorado beetle multiplication and spread, as 21 outbreaks were found and eradicated in the Thames estuary area during 1933 and 1934. There is also evidence of an influx of beetles to the same area in the autumn of 1935, which did not lead to establishment of colonies. In fact the risk was considered so great that inspections of potato crops were made in coastal areas from Harwich to Southampton late in the summers of 1939 and 1940, but no beetles were found (Bartlett, 1979). At the beginning of the 21st century the UK still remains free from this pest.

The War Years, 1939–1945

As might be expected, few new plant health initiatives were started during the years of the Second World War. However, in The Netherlands some seed growers and merchants had become dissatisfied with arrangements provided by the NAK and set up a separate quality inspection service of their own. This prompted the government to establish the Nederlandse Algemene Keuringsdienst voor Groente-en bloemzaden (NAKG) in 1941 to cover quality inspections of vegetable and flower seeds. Quality inspections for vegetative horticultural planting material had been started as a separate department of the NAK in 1935, for which task Mr Martinus Erkelens was recruited. Due partly to his energy and enthusiasm, and partly to demands from growers of planting material, this work expanded rapidly. In 1943 the horticultural department of the NAK was abolished and responsibility for quality inspections of tree crops, strawberry runners and other fruit planting material was taken over by the newly established Nederlandse Algemene Keuringsdienst voor Boomwekerijgewassen (NAKB; Dutch Inspection Service for Tree Crops). The NAKB developed under the guidance of Mr Erkelens, who became director from 1946 until his retirement 20 years later (Anon., 1993, 2000).

In the UK, the need to safeguard and increase home-grown food supplies necessitated review and modification of existing measures and the adoption of a few new ones. A new Wart Disease of Potatoes Order in 1941 discontinued designation of Infected Areas and Clean Land, but maintained the prohibition of planting susceptible varieties on infested land. It also established a Protected Area in the heartland of potato production in the east of England in the areas bordering the Wash, aimed at protecting the production and export of the very popular but susceptible variety, King Edward.

Verticillium wilt of hops

The severe or 'progressive' strain of the hop wilt disease, caused by the fungus *Verticillium albo-atrum*, was described in 1933 from Kent, south-east England, and had spread sufficiently to cause alarm in the English hop industry in the years just before the outbreak of war. Recognizing that much spread was due to the planting of new hop gardens with infected cuttings taken from mother plants in gardens where the disease was known to occur, a voluntary Scheme for the Inspection and Certification of Hop Gardens for freedom from verticillium wilt was inaugurated by the Ministry of Agriculture in 1943.

Post-war Developments in the UK

Administrative control of hop wilt

The voluntary scheme for inspection of hop gardens was soon supported by legislation. The Progressive Verticillium Wilt of Hops Order, 1947, made the disease notifiable and prohibited the sale of hop plants from land known to be affected by the disease. By 1950 many propagators had established special permanent layer beds to provide planting material, and had also started to produce plants by means of softwood cuttings, but the programmes for breeding wilt-resistant varieties were as yet only in the early stages of development. The A-plus Scheme for hops was introduced in 1955 for wilt-free nurseries and established a certification system for rooted plants propagated from wilt-free stock. Virus-tested clones of the major varieties were produced at East Malling Research Station, Kent, from 1966, and from 1975 the AA grade of certification was awarded to plants propagated from this new pathogen-tested stock, while certification of hop gardens finally ceased.

Certification schemes for fruit trees

Work on virus diseases of tree fruits progressed rapidly after the war in both Europe and America, notably at the British research stations of East Malling and Long Ashton. Schemes for the certification of fruit plants that were free of virus diseases were extended to tree fruits, starting with a scheme in England for certification of vegetatively propagated apple and plum rootstocks in 1946. Only visual health checks were made, but trueness to type was authenticated and buyers had some assurance of receiving fruit trees on correctly named, unmixed and uniform rootstocks that would behave predictably in the orchard. Quince and cherry rootstocks followed in 1954. By 1955 a considerable number of virus-tested clones of tree fruits and their rootstocks were available. Some of these clones, such as the M.9a apple rootstock, still contained latent virus infections, which were not removed until heat therapy and shoot tip grafting techniques were applied to tree fruit from 1956 onwards. At the East Malling Research Station a system of distribution of this material to members of The Kent Incorporated Society for Promoting Experiments in Horticulture (under the auspices of which the Research Station originated) was started with the release of four virus-tested cherry cultivars in 1953. This was known as The Mother Tree Scheme, although rootstocks were also distributed. Because the work had been done at the East Malling and Long Ashton research stations, the name EMLA was coined in 1965 to cover the resulting pathogen-tested nuclear stock of the highest health status. The objective of the EMLA initiative was to issue a collection of clones of commercial scion and rootstock varieties, each one of which would be free of all known viruses and of proven trueness to variety and agronomic performance. Release of these clones as a collection would make it worthwhile for nurserymen to maintain them in a separate nursery with adequate isolation from their ordinary stock. The first release of EMLA material (the cherry F12/1 rootstock) was made in 1968.

Colorado beetle control in England

A total of 20 Colorado beetle colonies were found in England during the war years, all of which were eradicated successfully. However, in the early post-war period, from 1946 to 1952, 119 breeding colonies were found. Eradication was again achieved, using both the old arsenical dusts and the new DDT insecticide sprays, and precautionary treatments were applied over a wide area in Kent, Surrey and to the west and north of London. No further breeding colonies were found until 1976, when one was found and eradicated in Kent, and a subsidiary colony in 1977. Beetles continued to be discovered each year, however, carried in or on all manner of substrates and produce, and in the post-war years interceptions on imported vegetables became increasingly frequent. This was doubtless due to the increased importation of vegetable produce from infested parts of Europe and elsewhere, where vegetables grown on land that had previously carried an infested potato crop tended to harbour beetles emerging from diapause in the soil. Precautionary legislation was tightened after 1945, requiring inspection of produce before export, maintaining a modified radius freedom from infested areas for imports during the growing season, yet allowing imports from designated areas where active Colorado beetle control measures were in force. These measures were costly and cumbersome to administer, and from the late 1950s were replaced, for certain areas, by the design and establishment of approved packing stations, which minimized the risk of contamination. Existing measures were consolidated in the Importation of Plants, Plant Produce and Potatoes (Health) (Great Britain) Order, 1971.

Anglo-French Colorado beetle controls

In the early post-war years, weather conditions were generally favourable for Colorado beetle (especially the warm summer of 1947), which greatly increased the threat of its establishment, both in Great Britain and in the Channel Islands, where it had just been eradicated in Jersey. To counter this threat to the Channel Islands a containment campaign with statutory support was started in the Cotentin Peninsula of France under the auspices of the newly formed EPPO (see below). This campaign was later extended to the coastal areas of the French départments of Pas de Calais and Nord to protect the British mainland, and still continues, with costs shared between France, the UK and the Channel Islands. In addition, contingency plans are held in readiness by the British agricultural authorities, including the Central Science Laboratory, to deal with any Colorado beetle outbreaks that may occur (Bartlett, 1979).

Formation of the European and Mediterranean Plant Protection Organization

The disruption to organized Colorado beetle control during the Second World War resulted in large populations of beetles on the continent of Europe by the end of hostilities. After the war, European efforts to control Colorado beetle were coordinated by the International Committee for the Control of Colorado Beetle. The increasing risk of spread of pests in the rapidly expanding international trade of the post-war period also led to the establishment of the European Working Party on Infestation Control. These two committees had much in common and eventually were combined to form EPPO, which was established in April 1951 with its headquarters in Paris (Smith, 1979). Although it predated the IPPC by a few months, it became the first RPPO to operate under the IPPC (see Chapter 3).

International Phytosanitary Controls*

Introduction

International phytosanitary activities today are governed by relatively few agreements and organizations, principal among which are the Agreement on the Application of Sanitary and Phytosanitary Measures under the World Trade Organization General Agreement on Tariffs and Trade (the WTO-SPS), the IPPC, administered by a Commission on Phytosanitary Measures under the United Nations Food and Agriculture Organization (FAO) and, more recently, the Convention on Biological Diversity (CBD) administered under the United Nations Environment Programme (UNEP).

Prior to adoption of the 1997 revision of the IPPC (see below) by the requisite two-thirds of contracting parties, an Interim Commission on Phytosanitary Measures (ICPM) was established to administer it, pending agreement on the details of such a commission. The IPPC governs the development and adoption of the International Standards for Phytosanitary Measures (ISPMs), which are referred to by the WTO-SPS and satisfy its criteria for international standards, but are established under the terms of the IPPC. The scope of the CBD includes alien species (especially those that are considered invasive) and genetically modified organisms and, therefore, where such organisms may be considered to be plant pests, it overlaps or interacts with the scope of the IPPC. Other international bodies important in phytosanitary affairs are the RPPOs, which relate to and operate under the IPPC and, for Europe, the Commission of the European Union, which coordinates phytosanitary

*By D.L. Ebbels and A.W. Pemberton.

activities within and on behalf of member states of the European Union. There are also bilateral and multilateral agreements between individual governments, which may affect their international phytosanitary activities directly or indirectly. This chapter deals with the IPPC, the RPPOs, the WTO-SPS, the CBD, and certain other agreements, while Chapter 4 covers the European Union phytosanitary regime.

Development of international cooperation

As described in Chapter 2, the devastation of vineyards and the wine industry in France and other European nations by the American vine louse (*Viteus vitifolii*, formerly *Phylloxera vastatrix*) was the impetus for the first international agreement to control a plant pest. This was the 1878 International Convention on Measures to be taken against *Phylloxera vastatrix* (Appendix II), which embodied many of the principles that are recognized today in international plant health.

The depressed state of agriculture in Europe in the years before the First World War, and the sudden need to increase agricultural production in the combatant countries, emphasized these and other agricultural problems and strengthened the role of the IIA, which had been established in Rome in 1905 (Chapter 2). International conferences on plant pathology and on plant protection hosted by the International Institute in 1914 (Anon., 1914; Rogers, 1914) and 1929 resulted in the International Convention for the Protection of Plants, 1929. However, this Convention was overtaken by the deteriorating international situation and subsequently the Second World War, so that it was ratified by only 12 of the 46 signatory countries, although another five adopted its principles without ratification (Chock, 1979).

The Food and Agriculture Organization and the International Plant Protection Convention

The Food and Agriculture Organization

The IIA was succeeded in 1946 by the FAO, with its headquarters based in the former Italian Ministry of External Affairs building in the Viale delle Terme di Caracalla in Rome (Fig. 3.1). It has developed enormously since then, and among its many important functions are the gathering, assessing and disseminating of information on agriculture and food, advising governments on action, and providing a neutral forum for discussion. FAO operates many major programmes of worldwide significance in the field of plant protection. Within the Plant Production

Fig. 3.1. Aerial view of the FAO headquarters buildings, Rome. Note the complicated design and that connections between the front building and the block behind are limited to certain floors. (Photo: FAO/Arma dei Carabinieri.)

and Protection Division, the Plant Protection Service supports many projects dealing with pesticides, plant pest control and related matters, particularly in developing countries. It also provides support for the IPPC and coordinates international control of migratory pests (Chapter 7).

The International Plant Protection Convention

Between 1946 and 1950 various representations on international plant protection had been submitted to FAO by its member countries. In May 1950, a draft plant protection agreement, which was largely based on a synthesis of these contributions, was presented at a conference convened jointly in The Hague by FAO and the government of The Netherlands. This was accepted and passed for detailed consideration to a panel consisting of plant protection officers from the USA and Canada who, with help from various other authorities, produced a revised version for consideration by a special session of the FAO Conference later the same year. More time to consider the draft Convention was requested by some countries and comments received were circulated to member countries by the FAO. A meeting of plant protection specialists in September 1951 combined and coordinated comments and responses and prepared a final draft of what became the IPPC, which was approved by the FAO

Conference of November 1951 (Ling, 1953). The IPPC was revised in 1977 and again in 1997.

From its inception, the IPPC was supported by FAO through an informal *ad hoc* secretariat within the Plant Protection Service. *Ad hoc* technical working groups of interested countries and, in recent years, technical consultations between the RPPOs (see below) were organized. Matters identified in these meetings could be passed through national governments to the FAO Committee on Agriculture (CoAg, normally meeting in May) for consideration as an amendment to the IPPC. With approval, such amendments were passed to the FAO Council (meeting in June) and subsequently to the biennially convened FAO Conference (in November) for formal adoption. However, with the Agreement on the Application of Sanitary and Phytosanitary Measures in the Final Act of the Uruguay Round of the WTO Multilateral Trade Negotiations, signed in Marrakesh on 15 April 1994 (see below), a need for ISPMs was identified. To develop these, the FAO Conference of 1993 established a Committee of Experts on Phytosanitary Measures and a Standard Setting Procedure. Several ISPMs were adopted under this system until the FAO Conference of 1997 agreed, as part of a major revision of the IPPC, that future ISPMs would be considered and adopted by the ICPM, which (after acceptance of the IPPC revision by two-thirds of the contracting parties) would become the Commission on Phytosanitary Measures (CPM). Approval of ISPMs via CoAg and the FAO Conference was thus superseded.

The 1997 revision of the IPPC was substantial, with the objectives of: (i) bringing it up to date with current phytosanitary practices; (ii) bringing it into line with the new concepts of the WTO-SPS Agreement (see below); (iii) establishing a new mechanism, the CPM, for the formal setting of the phytosanitary standards that would be recognized under the SPS Agreement; and (iv) formally establishing a new IPPC Secretariat. The IPPC has a common interest with the SPS Agreement in that it covers the application of phytosanitary measures affecting international trade. However, it is distinct in having its own scope and objectives oriented towards plant protection rather than trade. One notable aspect of the revision was the extension of the IPPC definition of 'plants' to cover forests and wild flora, thus clarifying the use of phytosanitary measures to safeguard non-commercial plants, which formerly were not specifically covered. As it was agreed that the revision did not impose new obligations, the revised version of the IPPC comes into force when accepted by two-thirds of the contracting parties.

The International Plant Protection Convention 1997

The IPPC 1997 sets out the phytosanitary rules and policies to be applied by contracting parties (signatory governments) to limit the spread of pests. It aims primarily to combat the introduction and spread of quarantine pests in international trade, and it emphasizes: (i) the necessity for international cooperation; (ii) that phytosanitary measures should be technically justified; (iii) that their details and conditions should be available to all (i.e. should be 'transparent'); and (iv) that they should not be applied in such a way as to constitute a means of arbitrary or unjustified discrimination or a disguised restriction on trade. A summary of the main provisions follows, but the original and up-to-date legal text must be consulted for definitive requirements and detailed interpretation.

1. The IPPC establishes the principle that all countries have a joint responsibility in plant quarantine to adhere to the rules agreed, without prejudice to obligations under other international agreements (Articles I and III).

2. Each contracting party is required to establish an official national plant protection organization which can satisfactorily perform the required measures and administer the IPPC provisions. A description of the organization must be submitted to the IPPC Secretary and, on request, to other contracting parties. Duties specified include the issue of phytosanitary certificates; surveillance of plants (for detection of pests) in cultivation, storage, in transportation or in the wild; inspection and, if necessary, treatment to meet phytosanitary requirements, of regulated items moving in international trade; the protection of areas open to the establishment of a quarantine pest; performing pest risk analyses; ensuring the integrity of regulated consignments between phytosanitary certification and export; and the training of staff. In addition signatory governments must 'to the best of their ability' issue their phytosanitary regulations, publicize information concerning regulated pests and their control, and provide for research on relevant problems (Article IV).

3. Phytosanitary certificates must be issued as required to cover exports of regulated plants, plant products and other articles, following the format of the model certificate provided in the Annex to the IPPC. Signatory governments undertake not to require certificates of different format, but certificates in electronic form may be issued if acceptable to the importing country. Inspections and related activities are to be done by technically qualified public officials acting under the authorization of the official national plant protection organization, and with access to relevant information (e.g. the phytosanitary regulations of the importing country) (Article V).

4. National phytosanitary regulations are required to be embodied in national law and published, so that all can be aware of their requirements. Such regulations must be both the minimum necessary to achieve their purpose and technically justified on phytosanitary grounds. All phytosanitary measures must be limited to what can be technically justified. Those measures against pests that occur within the territory of the contracting party must be applied equally to both imports and domestic production. Measures must be reviewed as necessary, and cannot be required against pests whose potential impact has not been shown to be economically unacceptable (Articles VI and VII).

5. Nation states have the right to adopt phytosanitary measures covering imports of plants, plant products and other relevant items. To prevent the introduction and/or spread of regulated pests, these may require such measures as inspection, refusal of entry, re-export, treatment, detention, restriction of movement, or destruction. This also applies to the regulated pests themselves and to biological control agents or other organisms of phytosanitary concern claimed to be beneficial. In addition, specified points of entry must be such as not unnecessarily to impede international trade and these must be notified to the RPPOs of which they are a member, and others affected. Phytosanitary measures may be applied to consignments in transit only where these are necessary and can be technically justified. Emergency action may be taken to combat a potential pest; this must be reported to their RPPO and continuing action must be evaluated to ensure it is justified (Article VII).

6. Each contracting party must designate a contact point for exchange of information, and is obliged to cooperate as fully as possible in providing information on, and reporting the occurrence or interception of, plant pests, and to participate in any international action to meet phytosanitary emergencies. Cooperation is also obligatory for establishing RPPOs and in the development of international standards for phytosanitary measures. These should be taken into account when formulating phytosanitary requirements or taking phytosanitary action (Articles VIII, IX and X).

7. The IPPC provides an impartial forum for the settlement of disputes. In this respect, if the parties cannot resolve the matter among themselves, the Director-General of FAO can be requested to appoint a committee of experts to examine and report on the disputed matter. Such a report is not binding, but should be used as the basis for further consideration of the problem and may also be submitted to the appropriate international organization for resolving trade disputes (which will normally be the WTO SPS Committee) (Article XIII).

8. Articles XI and XII provide for the establishment of a Commission on Phytosanitary Measures (CPM) within the framework of the FAO and for a Secretariat to administer it. The objectives, rules and duties of the CPM and the Secretary are specified.

9. Articles XIV to XXIII deal with substitution of prior agreements (but not the *Phylloxera* convention of 1878), territorial application, supplementary agreements, ratification and adherence, non-contracting

parties, languages, technical assistance, amendment, entry into force and denunciation.

Regional Plant Protection Organizations

Intra-regional cooperation with the aim of forming RPPOs is required under Article IX of the IPPC, and inter-regional cooperation via these RPPOs is expected. The RPPOs are required to cooperate with the IPPC Secretariat to achieve the objectives of the IPPC and, in turn, the CPM Secretary must convene regular technical consultations for RPPOs, promote ISPMs and encourage inter-regional cooperation.

RPPOs now cover most areas of the world and all countries have the opportunity to become members of an RPPO. Summaries of information on the current RPPOs are given in Appendix I. The main functions of RPPOs are to act as phytosanitary coordinators for their respective regions, to gather and disseminate information, to promote harmonization of phytosanitary regulations, to encourage the adoption of sound phytosanitary policies, and to promote the objectives of the IPPC. However, RPPOs have no legal force and have advisory powers only, although these can be strong. Membership of RPPOs is voluntary and some of the areas covered by the RPPOs overlap, so that a country may belong to more than one RPPO. Some RPPOs are financed by subscription from the member countries, the amount of subscription often being in relation to the size of the country or its economy. However, the FAO supports some RPPOs by providing secretariat assistance and facilities at FAO regional offices. Some RPPOs have been dissolved or modified and others formed, so that the situation is not static and changes gradually with these developments over the years. There is also a wide variation in the size and format of RPPO operations, some being much more active than others. To date, two of the most active have been the North American Plant Protection Organization (NAPPO) and European and Mediterranean Plant Protection Organization which, for this reason, are covered in more detail in the information given in Appendix I. All have small or very small secretariats and sometimes the secretariat rotates between the member countries. Under the auspices of the IPPC, the annual technical consultations between the various RPPOs can be convened in any of the different regions. These aim to develop the objectives of the IPPC, coordinate activities of mutual interest, discuss mutual problems, and make recommendations to the IPPC Secretariat. The RPPOs currently in operation are described briefly in Appendix I and the regions they cover are shown in Fig. 3.2. Up-to-date information on these topics can be found on the website www.ippc.int/cds_ippc/IPP/En/default.htm

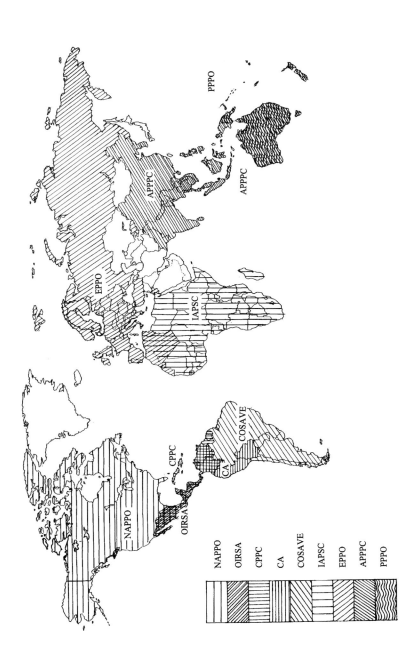

Fig. 3.2. Areas of the world covered by the Regional Plant Protection Organizations. NAPPO, North American Plant Protection Organization; OIRSA, Organismo Internacional Regional de Sanidad Agropecuaria; CPPC, Caribbean Plant Protection Commission; CA, Comunidad Andina; COSAVE, Comité Regional de Sanidad Vegetal del Cono Sur; IAPSC, Inter-African Phytosanitary Council; EPPO, European and Mediterranean Plant Protection Organization; APPPC, Asia and Pacific Plant Protection Commission; PPPO, Pacific Plant Protection Organization. (Diagram: Catherine Ebbels.)

International Standards for Phytosanitary Measures

Since the establishment of the IPPC standard setting procedure in 1993, several ISPMs (or their revisions) have been adopted each year, giving a current total of 19 (Table 3.1). The programme for the development of ISPMs is managed by the ICPM on the basis of the guidance of the ICPM Standards Committee and Secretariat. An expert working group (normally of five to six specialists) is established to prepare an initial draft, which is then further refined by the Standards Committee, following at least one written consultation with all IPPC contracting

Table 3.1. International Standards for Phytosanitary Measures (ISPMs).

ISPM No.	Title	Published
ISPM 1	Principles of Plant Quarantine as Related to International Trade	1995
ISPM 2	Import Regulations. Guidelines for Pest Risk Analysis	1996
ISPM 3	Import Regulations. Code of Conduct for the Import and Release of Exotic Biological Control Agents	1996
ISPM 4	Pest Surveillance. Requirements for the Establishment of Pest Free Areas	1996
ISPM 5	Glossary of Phytosanitary Terms, 2003 (including Supplement No. 1: Guidelines on the Interpretation and Application of the Concept of Official Control for Regulated Pests, 2002 and Supplement No. 2: Guidelines on the Understanding of Potential Economic Importance and Related Terms Including Reference to Environmental Considerations, 2003)	2003
ISPM 6	Guidelines for Surveillance	1997
ISPM 7	Export Certification System	1997
ISPM 8	Determination of Pest Status in an Area	1998
ISPM 9	Guidelines for Pest Eradication Programmes	1998
ISPM 10	Requirements for the Establishment of Pest Free Places of Production and Pest Free Production Sites	1999
ISPM 11	Pest Risk Analysis for Quarantine Pests	2001
	Supplement No. 1: Analysis of Environmental Risks	2003
ISPM 12	Guidelines for Phytosanitary Certificates	2001
ISPM 13	Guidelines for the Notification of Non-compliance and Emergency Action	2001
ISPM 14	The Use of Integrated Measures in a Systems Approach for Pest Risk Management	2002
ISPM 15	Guidelines for Regulating Wood Packaging Material in International Trade	2003
ISPM 16	Regulated Non-quarantine Pests: Concept and Application	2002
ISPM 17	Pest Reporting	2002
ISPM 18	Guidelines on Lists of Regulated Pests	2003
ISPM 19	Guidelines for the Use of Irradiation as a Phytosanitary Measure	2003

parties. The process from inception to adoption normally takes about 2 years, but complex standards can take much longer and involve several rounds of expert development and contracting party consultation. The standards are science based and the fundamental issues are rarely considered controversial. Adoption is therefore normally a simple procedure. In certain cases, however, the science is not yet developed but the international community still requires guidance. Under these circumstances compromises must be reached to enable final decisions to be taken by the ICPM.

The ISPM programme aims to develop a series of largely interlinked standards. The majority of those adopted in the early years consist of 'horizontal' standards providing general guidance for the development and operation of plant health procedures. With this framework in place, 'vertical' pest- or commodity-specific standards are being developed to give precise guidance for dealing with particular situations.

Of the 'horizontal' general standards, by far the most important is ISPM No. 1 (Anon., 1995), *Principles of Plant Quarantine as Related to International Trade*. Its aim is to facilitate the development of other standards and to guide phytosanitary authorities so as to reduce or eliminate unjustifiable phytosanitary measures. Its purpose is to support the IPPC and related agreements and to ensure coherence with the WTO-SPS. It specifies general principles, which should be considered together as a single entity, and other more specific principles related to particular procedures.

The general principles concern sovereignty (the right to take measures), necessity (that measures taken must be due only to phytosanitary considerations), and minimal impact (the measures must be the least restrictive available to achieve the phytosanitary objective). They also concern modification of measures (which may need to change as new facts become available), transparency (the publication and provision of information), harmonization (based on international standards, if available) and equivalence (acceptance of alternative measures if they have the same effect). The final general principle relates to dispute settlement, where resolution should be found at the technical bilateral level, formal procedures only being used as a last resort.

Specific principles concern technical authority (provision of an official Plant Health Service), risk analysis (see Chapter 11), managed risk (there is always some risk, so measures must be aimed at managing the risk defined), pest free areas (which must be utilized where available), non-discrimination (no discrimination between countries or between imports and domestic material in the same phytosanitary situation), emergency action (which can be taken even when only preliminary data are available, but which must be reviewed when further information is obtained), cooperation and notification of non-compliance.

The other 'horizontal' general standards vary in the extent of their applicability to plant health work. Those of widest application include ISPM No. 5 (Anon., 2002a), *Glossary of Phytosanitary Terms*. This ISPM is updated annually as new terms and definitions are established or require modification with the adoption of new ISPMs, and also contains guidance on the interpretation or application of terms such as that of 'official control'. Also of wide application are those on pest risk analysis (PRA), which forms the basis for regulations and for which ISPM No. 2 gives general guidelines and No. 11 focuses on PRA for quarantine pests (Anon., 1996b, 2001a). These are regularly updated as new techniques evolve or wider considerations are needed, such as to cover environmental risk or for the relatively new concept of regulated non-quarantine pests.

More specialized 'horizontal' standards include surveillance and eradication (ISPMs Nos 6 and 9; Anon., 1997a, 1998b), the establishment of pest free areas and pest free places of production (ISPMs Nos. 4 and 10; Anon., 1996d, 1999b), export certification and phytosanitary certificates (ISPMs Nos 7 and 12; Anon., 1997b, 2001b), pest reporting (ISPM No. 17; Anon., 2002d), and non-compliance notification (ISPM No. 13; Anon., 2001c).

Taken together, these 'horizontal' general standards provide guidance, interpretation, instruction and the approach that national plant protection organizations should adopt to dispense their obligations under the IPPC. International standards setting down specific instruction regarding the action necessary to deal with particular pests, individual techniques or technologies, or specific commodities are only just being developed. ISPM No. 3 (Anon., 1996c) provides a *Code of Conduct for the Import and Release of Exotic Biological Control Agents*. ISPM No. 15 concerns *Guidelines for Wood Packaging Material*, required to prevent the spread of a wide range of tree pests of quarantine concern, including *Bursaphalenchus xylophilus* (pine wood nematode) and *Anoplophora glabripennis* (Asian long-horned beetle). Examples of specific ISPMs are *Guidelines on the Use of Irradiation as a Phytosanitary Measure* and *Guidelines for the Surveillance of Citrus Canker*, which is in the process of development.

All current ISPMs are published on the IPPC home website, the International Phytosanitary Portal (IPP), at www.ippc.int/cds_ippc/IPP/En/default.htm and are available in the official languages of the FAO.

The World Trade Organization

Discussions held between 23 countries in 1946–1947 aimed at mutual reduction in tariffs on trade in goods and at creating an international trade organization as an agency of the United Nations. While the latter objective was not achieved, a General Agreement on Tariffs and Trade (GATT 1947) was reached, which came into force in January 1948. This covered a large number of tariff concessions and also laid down some ground rules for international trade. An *ad hoc* organization (also known as GATT), to provide a secretariat for supporting the implementation of the legal framework, was operated until it was replaced on 1 January 1995 by the World Trade Organization (WTO), based in Geneva, Switzerland. The WTO, a permanent and legally established international organization with about 500 staff, administers an updated version of GATT (GATT 1994), the General Agreement on Trade in Services, the agreement on Trade-Related Aspects of Intellectual Property Rights, and a dispute settlement system. Its main objectives are to encourage the flow of trade and to provide a forum for negotiations and dispute settlement. Detailed information is available on the WTO website: www.wto.org

The Agreement on the Application of Sanitary and Phytosanitary Measures

The Uruguay Round of negotiations under the old GATT administration took place between 1986 and 1994. Concluded in Marrakesh, Morocco, on 15 April 1994, it resulted in about 60 agreements, decisions and understandings. The Agreement on the Application of Sanitary and Phytosanitary Measures (the SPS Agreement) is one of these, which form part of the treaty establishing the WTO. All signatories to the WTO treaty are therefore Members of the SPS Agreement. It covers measures on food safety and preserving animal and plant health (including wild fauna and flora). In these matters it links to the Codex Alimentarius Commission (for standards, guidelines and recommendations on food additives, pesticide and veterinary drug residues, contaminants and methods of investigation), the International Office of Epizootics (for standards, guidelines and recommendations on animal health) and the IPPC (for standards, guidelines and recommendations on plant health). Within the WTO all matters relating to the SPS Agreement, except dispute settlement, are dealt with by the SPS Committee. Settlement of trade disputes involving SPS measures is handled under the WTO dispute settlement procedures. A short account of the WTO-SPS and its application to plant health and quarantine is given by McRae and Wilson (2002).

As tariffs and other trade barriers are reduced under the WTO agreements, governments are often likely to be pressured to protect domestic production from foreign economic competition by using measures ostensibly designed to ensure food is safe to consume or to protect animals or plants from exotic pests, but which actually go beyond what is necessary or reasonable for this purpose and constitute a barrier to trade. Such measures can deceptively exercise very effective covert control on trade, while being very difficult to challenge because of their highly technical nature. The main aim of the SPS Agreement is therefore to prevent the abuse of health protection measures for trade protectionist purposes, while maintaining the right of governments to take necessary and justifiable measures to maintain the level of health protection it considers to be appropriate. It also requires a consistency of approach to decisions on SPS matters. A summary of the main points of the WTO-SPS follows, but the original and up-to-date legal text must be consulted for definitive requirements and detailed interpretation.

1. Sanitary and phytosanitary measures must be applied only for protecting food safety or animal or plant health, and the measures must be shown to be justified on objective scientific grounds to meet the threatening risk. Another aim of the SPS Agreement is to encourage uniform treatment of trade and to prevent unjustified discrimination, both between domestic and foreign products or traders, and between different foreign trade partners. Where trading partners can show that different measures provide an equivalent level of protection, these should be accepted. SPS measures should be the minimum necessary to achieve the appropriate level of health protection (Articles 2, 4, 5 and Annex C).

2. The WTO-SPS regards measures which conform to internationally agreed standards, where these exist, as being in compliance with the Agreement. In the phytosanitary field, therefore, this links to the ISPMs developed and adopted under the IPPC but not to regional standards developed and adopted under the auspices of RPPOs. Although conformation to international standards is not obligatory, governments may be asked to provide scientific justification where national measures are more restrictive of trade (Article 3).

3. It requires that the scientific process of pest risk analysis (PRA) is used to determine what measures are appropriate to the circumstances, and it specifies which factors should be considered. When a national measure is felt to be scientifically unjustified and is restrictive to trade, the affected member may request an explanation of the basis for the measure, which must be provided (Article 5).

4. Measures must be appropriate to the conditions in both the area of production and destination, and take into account the prevalence of pests, including the existence of pest free or low pest prevalence areas, evidence for which must be provided and reasonable access for verification allowed to trade partners on request (Article 6).

5. National regulations must be published promptly, allowing time for trading partners to comply, and must be notified to the WTO

Secretariat. Governments must each establish an Enquiry Point to disseminate information on their SPS measures and must designate a single central government authority responsible for notification procedures. The basis of SPS measures must be made available for scrutiny, if requested. Publication of phytosanitary measures, and making information on them available to contracting parties, is known as making them 'transparent' and is a key objective of the Agreement (Article 7, and Annex B).

6. Procedures for control, inspection and approval must be non-discriminatory, and requirements for information, sampling and facilities must be limited to what is necessary and appropriate for such procedures, and a complaints procedure must be operated (Article 8 and Annex C).

7. The special difficulties and needs of developing countries are specifically addressed by encouraging or requiring provision of technical assistance and special consideration (such as extended time for compliance) from developed countries to less-developed trading partners. Implementation of SPS provisions may be delayed by developing countries, and the SPS Committee is also able to grant time-limited exceptions from SPS Agreement obligations to developing countries (Articles 9, 10 and 14).

8. The establishment and operational rules of the SPS Committee are laid down and members are required to ensure that regional or non-governmental bodies within their territories also comply with the provisions of the Agreement (Articles 12 and 13).

The emphasis on scientific analysis and justification has significantly raised the profile of the practice of PRA, for which ISPMs Nos 2 and 11 have been developed. PRA is a process which (in the phytosanitary field) seeks scientifically and objectively to analyse phytosanitary risks and thus to determine what countermeasures may be justified. Similarly, it can be used to analyse phytosanitary measures to determine whether they are necessary and justifiable (see Chapter 11).

The complete freedom of governments to designate 'the appropriate level of protection' they wish for the safety of food or the health of their animals or plants, as defined in the WTO-SPS Annex A (5), has been keenly debated. In theory, governments could set this at such a high level that it would make international trade very difficult. However, there are constraints to this. First, a risk must be identified, assessed and measures applied only to the extent necessary to counter the risk (Articles 2.2 and 5.1). Under Article 5.4 the appropriate level of protection chosen should take into account the objective that there should be the minimum disruption to trade, and under Article 5.8 the rationale for measures not conforming to international standards must be provided on request. Second, under IPPC Article VII 2.g, measures must also be consistent with the pest risk involved, be the minimum necessary to meet the risk, and result in the minimum disruption to commerce and travel. In addition, the principle of non-discrimination (WTO-SPS Article 2.3, IPPC Article VI 1.a) requires that the same standards are also operated for domestic production. Nevertheless, all countries may demand very high levels of protection in certain circumstances, for example, when the importing country is free from a serious pest prevalent and not being controlled in the exporting country. However, the measures imposed must still be based on PRA and scientific argument. Their justification may be tested and, if necessary, challenged on the basis of PRA results.

Convention on Biological Diversity

Both the IPPC and WTO-SPS make reference to protecting wild plants and the environment, but these agreements are generally considered largely to concern trade. The CBD (UNEP, 1992) has the objective of the 'conservation of biological diversity and the sustainable use of its components' (Article 1). It recognizes that one of the major threats to diversity is the spread of 'alien species which threaten ecosystems, habitats and species' and requires contracting parties to prevent their introduction or control or eradicate them (Article 8 h). In addition, the CBD recognizes the potential threat of genetically modified organisms or 'living modified organisms resulting from biotechnology' (LMOs), to use the CBD terminology. These may have adverse environmental impacts that could affect the conservation and sustainable use of biological diversity, and contracting parties should therefore regulate, manage or control their use or release (Articles 8 g and 19.4).

To help governments meet their obligations, two protocols have been established under the CBD, the *Cartegena Protocol on Biosafety* (UNEP, 2000) and the *Guiding Principles for the Prevention, Introduction and Mitigation of Impacts of Alien Species* (UNEP, 2002).

The Biosafety Protocol

This is aimed at ensuring an adequate level of protection for the safe transfer, handling and use of LMOs, focusing in particular on transboundary (international) movements (Article 1 of the Protocol). The requirements are both detailed and complex. The precautionary approach and the responsibilities of the exporting country are stressed. At its heart is a notification procedure required of the exporting country, a risk analysis procedure, and an associated authorization procedure with set response times required of the importing country (Article 10 specifies 90 days to recognize received information and 270 days to complete its analysis and communicate a decision), although there is also the facility for both a simplified procedure and other arrangements, such as bilateral agreements (Articles 13 and 14). The annexes to the Protocol set down the requirements for information to be notified by the exporter (such as taxonomic status, intended use, a description of the modification, and any existing risk assessment) and the aspects that must be evaluated in the risk assessment (such as the identification of any genotypic or phenotypic characteristics, which may have adverse effects, the likelihood of these being realized, and the evaluation of the consequences).

The Guiding Principles on Alien Species

This establishes 15 guiding principles within a four-part framework covering general aspects, prevention, introduction of species and mitigation of impacts. In particular, Principle 1 endorses the CBD concept of a 'precautionary approach', where lack of scientific certainty should not be used as a reason for not taking action. A three-stage hierarchical approach (prevention, eradication, containment) based on the ecosystem is required, with the country of origin shouldering the main responsibility of preventing spread of alien invasive species to other countries (Principles 2, 3 and 4). Border controls and precautionary measures should be applied, with measures based on risk assessment (Principles 7 and 11). In case such measures should prove ineffective, surveillance systems are recommended to facilitate early detection of new species and, where detected, eradication or containment measures should be taken (Principles 5, 12 and 14). Control measures targeted at reducing damage (Principle 15) should be considered only as a last resort. For intentional introductions (such as for research, environmental amelioration or economic or industrial development, including agriculture or horticulture) an authorization procedure requiring risk assessment and the precautionary approach should be established (Principle 10). Other principles concern research, public awareness, cooperation, information exchange and capacity building (including assistance to developing countries). Further information is available on pages 247–252 of the website www.biodiv.org/doc/meetings/cop/cop-06/official/cop-06-20-en-pdf

Implications for plant health services

Many aspects of the CBD, its Guiding Principles and the Biosafety Protocol have far-reaching implications with consequences for plant health services. An alien species that is a plant pest (such as a pathogen or an invasive weed) and threatens ecosystems, habitats or species would be considered a quarantine pest under the IPPC, as also would an LMO which is a plant pest threatening biodiversity. Import controls and precautionary measures involving eradication, containment, prevention of spread, risk assessment, surveillance, inspection, import licensing and many other aspects are all routine tasks for plant health services, although restricted in scope to those aspects relating to plant pests and plant health. Neither the IPPC nor the CBD takes precedence over the other, and there is an obligation on contracting parties to respect both conventions. Plant health services and environmental agencies must therefore work together to ensure that conflicts of interest do not arise or are quickly resolved, that duplication of effort is avoided and that experience and expertise is shared to ensure efficient use of limited resources. Plant health expertise and systems can be used either to meet the obligations of the CBD, where appropriate, or as a basis for developing other specialist systems. The CBD and associated environmental expertise can also be utilized by plant health services to dispense their obligations under the IPPC and WTO-SPS to protect wild flora and the environment.

One of the more significant debates has concerned the application of the precautionary approach (Griffin, 2000). Because of the complexity of ecosystems, it could be envisaged that there will always be a lack of scientific certainty, thus triggering a reaction that all movement of species should be blocked and, by inference, so should all movement of goods and people that may act inadvertently as carriers. Such retrenchment of trade and travel would, of course, not receive the support of the general public. There is thus the necessity to identify potential hazards realistically, to estimate the probability of the hazards being realized and their potential consequences, to note areas of uncertainty while taking steps to address uncertainty where this is possible, and to formulate measures that can be justified technically to mitigate such hazards and which also take reasonable account of unresolved uncertainty. The respective competencies of the IPPC, CBD and other agreements at both the international and national levels can be utilized to help resolve such issues, with the long-term aim of such resolutions being incorporated in international standards, guidelines or protocols.

Other International Agreements Concerning Plant Pests

Regional agreements

Although RPPOs make recommendations to their member countries, these are not legally binding, although their members have a moral obligation to respect such recommendations. However, for some groups of countries, legally binding regional agreements have been reached.

In Europe, all member countries of the European Union have agreed to harmonize their plant health regulations into a single, commonly applied set of regulations relating to traded plants, plant products and other regulated articles. A series of regulations, of which the most important is the Council Directive 2000/29/EC, concern both imported commodities and commodities moved within and between member states. They also concern operational procedures, such as inspection, surveillance, reporting and control, with harmonization of procedures being facilitated by a Standing Committee on Plant Health and permanent European Commission staff, who include Community phytosanitary inspectors. These arrangements are described more fully in Chapter 4.

Other regional agreements are not so comprehensive. In South America the MERCOSUR group of countries (Argentina, Brazil, Paraguay and Uruguay) do not have harmonized regulations but agree that their national regulations respect common standards established under COSAVE, the RPPO for their region (see Appendix I). Such cooperation promotes increased confidence in their respective plant health systems and ensures that trade restrictions are minimal, thus facilitating freer trade between the MERCOSUR countries.

The North American Free Trade Agreement (NAFTA) also facilitates freer trade between its three member countries, Canada, Mexico and the USA. This is based on the agreement to respect international standards and North American regional standards (those developed by NAPPO, see Appendix I). Details are given on the websites www.nafta-sec-alena.org/ and www.sice.oas.org/trade/nafta/chap-073.asp#A709 NAPPO also offers technical support to the NAFTA-SPS Committee in the area of plant health. Thus, although the individual countries maintain their sovereign right to establish national plant health measures, the confidence established through their cooperative procedures means that, for trade between the member countries, these can be kept to a minimum.

Bilateral agreements

Although it is desirable for all countries to publish their regulations and respect the principles of transparency, minimal impact and non-discrimination, it is often the case that lack of information or resources impedes the establishment of comprehensive regulations to cater for all circumstances. In these situations an agreement between two countries may be established to allow trade under defined circumstances. Such agreements can be particularly beneficial to developing countries, and confer the major benefits of information exchange, development of trade, the testing of systems and the establishment of confidence. Provided that such agreements respect international obligations, particularly with regard to transparency and non-discrimination, then they are to be welcomed, especially if the eventual aim is for these agreements to be transposed into normal regulations, thus facilitating multilateral trade. Unfortunately, such considerations appear not always to be respected and it can be difficult for third parties to be fully informed on such agreements.

Controls on the use of plant pests as offensive agents

Certain international agreements, particularly the treaty commonly known as the Biological Weapons Convention 1972, prohibit the development, acquisition, retention or use of plant pests as offensive agents by contracting parties. This subject area does not fall within the scope of plant health and quarantine. However, those working in the plant health field may sometimes find themselves involved in dealing with matters related to this topic, particularly in providing information and giving advice, as the expertise and knowledge required is very similar to that needed for plant health. A brief account of the topic is therefore given in Appendix III.

The European Union Plant Health Regime   4

Introduction

Throughout history the countries of Europe have been bound together by numerous treaties and agreements. At first these tended to be bilateral, involving only the two participating countries but, as time went on, it became more common for agreements to involve more countries. After the Second World War, which was particularly intense in Europe, the idea of close cooperation between the countries of Europe became increasingly attractive, both for the perceived economic benefits and also to eliminate the prospect of hostilities recurring in the future. The plant health and quarantine regulations of the European Union have been one of the many results of this cooperation and are among the key features in the agriculture sector for adoption by countries aspiring to become Member States.

Milestones to the EU Single Market

The Treaty of Rome, signed in 1957, created the European Economic Community (EEC) of the six participating countries: Belgium, France, the Federal Republic of Germany, Italy, Luxembourg and The Netherlands. The emphasis at first was on economic cooperation and free trade between the participants, which were designated as *Member States*. In 1967 the councils and executive authorities (*commissions*) of the EEC, the European Coal and Steel Community, and the European Atomic Energy Community were merged, and the three communities thenceforth operated under a single council and a single commission.

In the succeeding years the tendency in common parlance to refer to this grouping as the 'European Community' gradually increased, EEC regulations began to cover much wider areas than strictly economic activities, and this development gradually accelerated. Plant health regulations featured in this development at an early stage and resulted in the directives of 1966–1969 on the marketing of field crop seeds, seed potatoes (66/403/EEC, now replaced), vine reproductive material (68/193/EEC), forestry reproductive material (66/401/EEC), and the control of potato wart disease (69/464/EEC) and of potato cyst nematode (69/465/EEC).

Throughout this period attention was also given to the development of plant health regulations covering trade in plants and plant products between Member States and between Member States and other countries (designated as *third countries*). This was temporarily interrupted by the accession of Denmark, Ireland and the UK in 1973. Development of plant health regulations was then resumed between members of the enlarged Community, and reached a conclusion in 1976 with agreement on the plant health directive (Council Directive 77/93/EEC). The original text of this directive went a long way towards harmonizing the diverse phytosanitary regulations of Member States, and established certain key principles, such as the transparency of regulations (no requirements hidden in unpublished documents), a common list of quarantine organisms and the right to take emergency action in certain circumstances. The International Phytosanitary Certificate and the rules governing its use still provided the basis for trade in plants and plant products between Member States. In order to achieve agreement, many of the different regulations of individual Member States were embodied in the Directive in a way that still applied only to the relevant Member State, causing the Directive to be extremely complicated and resulting in the perpetuation of considerable differences in phytosanitary requirements between Member States.

A common market was a political ideal supported by all Member States, but in the mid 1980s there still remained many barriers to the free flow of trade within the EEC. To compete in world markets, especially in the high technology sector, the Member States felt that it was essential to remove these obstacles and at the Milan meeting of the European Council (Heads of Government) in June 1985 agreement was reached on a timetable to achieve this goal. In 1985 the white paper *Completing the Single Market* was published and shortly afterwards the political decision to give effect to this was made with agreement on and signing of the Single European Act, 1985. This provided the basis for its implementation with a 7-year programme of legislation to remove some 300 identified barriers to the operation of a single internal market.

The political decision having been taken, the European Commission, in cooperation with the plant health services of the Member States, in effect had to go away and work out how best to protect plant health and control the spread of plant quarantine organisms while allowing trade to flow unhindered between the Member States. This initiated 7 long years of discussion and negotiation in the Council working groups and the standing committees. Several novel concepts were introduced (which are considered in more detail below), of which plant passports was probably the most innovative. Many of the phytosanitary conditions previously applying to only one or a few Member States were either abolished or were confirmed as applying to the whole Community. This resulted in a series of amendments to Directive 77/93/EEC, leading to the introduction of harmonized regulations scheduled to come into effect in all Member States on 1 January 1993. In practice, it proved impossible to adhere to this time schedule and the introduction of the Single Market plant health regime was postponed by 5 months. It came into force on 1 June 1993. Events leading up to the Single Market regime and its probable consequences were reviewed at a conference held at Reading, UK, in March 1993 (Ebbels, 1993).

Meanwhile, at the Maastricht meeting of the European Council in December 1991, agreement was reached on the Treaty on European Union. This included arrangements for monetary union and many other provisions for gradual progress towards other common policies. It also, at last, confirmed the new title of the EEC as simply the European Community (EC). Completion of ratification by all Member States brought the Treaty into force on 1 November 1993. Since this date, the communities, together with their associated bodies, in general have been known as the *European Union* (EU). Strictly speaking, within the EU the responsibility of the EC is limited to covering the core activities which come under the Council of Ministers, the European Commission and the other key Community institutions, while other activities of the EU (such as foreign and security policies) are organized on an intergovernmental basis.

Over the years, the number and complexity of amendments and directives associated with Directive 77/93/EEC increased to such an extent that in the year 2000 a consolidated version, Council Directive 2000/29/EC, was agreed. This new Plant Health Directive, together with its amendments, is now the main piece of legislation governing the EU plant health regime.

Information on the EU plant health regime

The political and administrative workings of the EU are both complicated and constantly changing. Excellent general guides to the EU have been produced, but quickly become out of date. The latest editions of the general introduction by Leonard (2000) and the guide to legislation by Foster (2000) are recommended. The EU has also developed a phraseology of its own, with numerous acronyms and terms that may not always have exactly the same meaning as native English speakers would expect. A large number of these are elucidated in the dictionary by Ramsey (2000), and those used here can be found in the Glossary. However, even the latest books will not contain very recent changes. For up-to-date information on all aspects of the EU, the best source is the EU website: http://www.europa.eu.int A general index and information on plant health legislation, including legislative texts, can be found on http://www.europa.eu.int/comm/food This also gives access links to reports on the deliberations of the various standing committees, forthcoming agendas, information on the work and structure of the scientific committees, and EU reports on country plant health inspections. EC legislation, announcements, calls for tenders or research proposals, and many other documents are published in the *Official Journal of the European Communities*.

Key EU Institutions

The administration of the EU and the institutions through which this is done have become increasingly complicated and more numerous. No attempt is made here to describe these in detail, and those requiring more information on EU institutions and procedures are referred to the sources quoted above. However, in order to explain the roles of certain key institutions that are directly involved in the creation and amendment of EC plant health legislation, simplified descriptions of these are given below. By this means, the basic procedures underlying EC legislation may also be appreciated.

The EU presidency

This rotates amongst the Member States at 6-month intervals. Although the order in which Member States held the presidency was formerly alphabetical, it is now arranged so that the presidencies of the larger states are more evenly spaced and so that states do not hold the presidency for the same half of the year on successive occasions.

The Council of Ministers

This is the senior political decision-making body. It is composed of representative ministers or officials from each Member State as appropriate for the subject area and level of operation. The European

Commission is also represented at the appropriate level. At the highest level it is composed of heads of government and the President of the Commission and is known as the *European Council*. This was formally confirmed by the Single European Act in December 1985. At lower levels the Council is composed of Ministers of State, junior ministers or officials of the ministry or department in each Member State as appropriate to deal with the agenda matter in hand. The Council may create sub-committees or working groups to deal with selected topics, and these are normally composed of appropriate civil servants and officials. Much of the detailed work is done in these committees. The chair for the Council and its sub-committees is taken by the representative from the Member State currently holding the EU Presidency. The Commission representative sits opposite the chairman, while the representatives from the Member States sit on either side in an order that is clockwise alphabetical in principle, but may diverge from this for various reasons mainly concerned with the rotation of the Presidency.

The European Commission

In some ways the Commission fulfils a similar role to that of the civil service in a sovereign state in that it is responsible for ensuring that the policies and legislation agreed upon by the Council of Ministers are implemented. However, it also has the role of initiating policies and legislative proposals. Except in certain particular circumstances, only the Commission may draft legislation; it is thus said that 'the Commission proposes and the Council disposes'. Thus the Commission has much greater powers and opportunities for innovation than an ordinary civil service. The Commission President and one or two commissioners from each Member State (comprising a current total of 20) are elected by the Council of Ministers. Each commissioner has responsibility for certain areas of government, and together they operate similarly to a cabinet of ministers. In the year 2000 the European Commission was reorganized, for purposes of administration, into 36 Directorates General (DGs) and Service Departments. The DGs, which are comparable to national ministries or departments dealing with particular areas of government, deal with areas of policy such as agriculture, environment, and various aspects of external relations, such as enlargement of the EU and trade. The Service Departments provide general and internal services, such as the Secretariat General, the Publications Office and the Legal Service. The commissioners are each responsible for one or more of these DGs and Service Departments. Those most often concerned with matters relating to plant health are:

- Enlargement: responsibility includes administration of programmes for assistance to countries wishing to become members of the EU (e.g. the Phare programme).
- Health and Consumer Protection: responsibility includes plant health and safety of plant protection products, and the Phytosanitary and Animal Health Inspectorates.
- Research: responsible for research policies, programmes and research funds.

The European Parliament

This is composed of Euro-MPs, representing constituencies throughout the EU. Its sittings alternate between Brussels and Strasbourg at 6-month intervals. The powers of the European Parliament are severely limited compared to that of a democratic sovereign state, but it is required to give opinions on all new draft legislation. If unfavourable, this normally results in amendment to the draft legislation, or at least a substantial delay and re-examination. By a system of committees, the European Parliament may also investigate problems and matters of interest.

Permanent Representation

All Member States maintain a delegation to the EU in Brussels, in much the same way as they maintain embassies or High Commissions in foreign states. This delegation is known as the Member State's Permanent Representation and is headed by a Permanent Representative who has ambassadorial status. Very close liaison is maintained between the governments of the Member States and their Permanent Representation. A committee of the Permanent Representatives, known as *COREPER* from the abbreviation of its French title, meets frequently (usually weekly) to consider all matters awaiting decision by the Council of Ministers. Decisions and proposals by Standing Committees and Council Working Parties therefore have to pass through COREPER before being considered by the Council.

The Economic and Social Committee

This committee and its standing sub-committees are composed of experts and lay persons nominated after consultation with national consumer and other interest groups. It is required to give opinions as to the economic and social effects of new draft legislation. If unfavourable, this can also result in amendment or delay and re-examination of the draft legislation. Although its powers are purely advisory, consultation of the Committee is normally routine and is mandatory in certain areas.

The European Court of Justice

Staffed by legal officers appointed by the Council, but with freedom for independent action, the Court is empowered to deal with all matters relating to the interpretation, execution or contravention of EC legislation.

EC Legislation

EC legislation may be formulated under resolutions of either the Council or the Commission. Council legislation normally represents the primary legislation, while the Commission legislation normally represents the secondary or implementing legislation. The main types of legislation met with in the plant health field are described briefly below. All these types can be used by both the Council and the Commission, with the title of each type including a designation indicating whether it derives from the Council or the Commission. Most of this legislation contains within it, at the beginning, a list of reasons as to why the legislation is necessary. These are presented in standard form, each generally beginning with 'whereas' and together they are known as the 'recitals'.

Regulations

These are directly binding, as written, on all Member States and there is no possibility for interpretation by individual Member States.

Directives

These are binding on all Member States and implementation is via national legislation. There is normally a time limit for such national legislation to be effected. A certain amount of interpretation is possible and is usually necessary in formulating the national legislation, which may also include national provisions not covered or required by the relevant Directive.

Decisions

These are selectively binding on certain Member States, according to the nature of the Decision.

Recommendations and Opinions

These are not binding upon Member States, but Member States are expected to consider and take note of these Recommendations and Opinions when dealing with matters to which they relate.

The Commission Standing Committees

Although many types of committees are formed *ad hoc* by the Commission to consider particular topics, it also maintains permanent committees, known as *standing committees*, to consider and deal with ongoing affairs in many areas of government. The standing committees are operated by the DG responsible for the particular area of policy. Much of the business of administering EU government in all spheres is done in these standing committees, and the plant health regime is no exception. In the margins of meetings they also provide a convenient forum for informal contacts on current issues between Member States. They are convened at fairly regular, sometimes frequent, intervals, but most regular EU business is suspended during the holiday seasons in the month of August and the weeks covering Christmas and the New Year. The meetings are chaired by Commission officials of the relevant DG. Representatives from each Member State are invited to participate, although attendance is not obligatory. As with Council meetings, representatives and delegates of Member States sit in clockwise alphabetical order.

The standing committees which are most often concerned with matters relating to plant health are described below, together with the principal legislation for which they are responsible. Many items have numerous amendments, which are not listed here.

Standing Committee on Plant Health

This committee was established by Decision 76/894/EEC and deals with both phytosanitary regulation and matters concerning the regulation of pesticides. The French title is Comité Phytosanitaire Permanent. It is administered by the Health and Consumer Protection DG and normally it is composed of different representatives and delegates, according to

Table 4.1. Principal phytosanitary legislation dealt with by the Standing Committee on Plant Health.

Commission Directive 2001/32/EC	Concerning protected zones
Council Directive 2000/29/EC	The Plant Health Directive; much the most important legislation dealt with by this standing committee
Commission Directive 98/22/EC	On minimum conditions at border inspection posts
Council Directive 98/57/EC	Control of potato brown rot
Commission Decision 97/647/EC	Interim diagnosis protocol for *Pseudomonas solanacearum*
Commission Regulation 2051/97/EC	Rules for EC financial contribution to plant health control
Commission Directive 95/44/EC	On scientific and trials derogations
Commission Directive 94/3/EC	On notification of interceptions
Commission Directive 93/50/EEC	Registration, producers of certain plants
Commission Directive 93/51/EEC	On movement in protected zones
Commission Directive 93/85/EEC	Control of potato ring rot
Council Directive 92/70/EEC	Surveys for protected zones
Commission Directive 92/90/EEC	Registration and obligations of producers and importers
Commission Directive 92/105/EEC	On plant passports
Council Directive 69/464/EEC	Control of potato wart disease
Council Directive 69/465/EEC	Control of potato cyst nematode

the subject matter with which the meeting is dealing. In the control of pesticides, it is concerned with all aspects of their registration, marketing and permitted residue levels, whereas with phytosanitary regulation it deals with conditions designed to prevent the spread of organisms harmful to plants, and related matters. This includes, for example, regulations relating to movement of specified commodities within and between Member States, imports from third countries, plant passports, listed and unlisted harmful organisms, interceptions, outbreaks and emergency procedures, protected zones and derogations (agreed exceptions to specified requirements). For phytosanitary matters the committee normally meets once or twice per month. The principal items of phytosanitary legislation dealt with are listed in Table 4.1.

Together with the EU Plant Health Inspectorate and the Commission Auditors, the Standing Committee on Plant Health also supervises crop protection programmes set up to aid the development of French and Portuguese overseas territories, which are part of the EU. These programmes, generally still referred to by their French and Portuguese acronyms POSEIDOM and POSEIMA, were established by Council Regulations EC 3763/91 and EC 1600/92 respectively, and provide finance to establish or improve plant health related projects and services in these territories (Guadeloupe, Guyana, Martinique, Réunion, the Azores and Madeira). One of the main aims in several of the territories is to reduce reliance on banana production.

Standing Committee on Agricultural, Horticultural and Forestry Seeds and Plants

Formerly known as the Standing Committee on Seeds and Propagating Material for Agriculture, Horticulture and Forestry, this standing committee deals with regulations relating to marketing and quality

Table 4.2. Principal legislation dealt with by the Standing Committee on Agricultural, Horticultural and Forestry Seeds and Plants.

Council Directive 2002/56/EC	On marketing of seed potatoes
Commission Regulation 1768/95	On farm-saved seed
Commission Regulation 1239/95	On Community plant variety rights (CPVR) proceedings and administration
Commission Regulation 1238/95	On CPVR fees
Council Regulation 2100/94	On CPVR
Commission Directive 93/62/EEC	Supervision of suppliers of young vegetable plants
Commission Directive 93/61/EEC	On standards for young vegetable plants
Commission Directive 93/17/EEC	On Community grades for seed potatoes
Commission Decision 92/231/EEC	On more stringent measures for seed potatoes
Council Directive 92/33/EEC	On marketing of young vegetable plants
Commission Directive 91/376/EEC	Seed of food and oil plants
Commission Decision 90/639/EEC	On derived vegetable variety names
Commission Directive 89/14/EEC	Isolation for vegetable beet seed crops
Commission Decision 87/309/EEC	Indelible printing on fodder seed packs
Commission Directive 86/109/EEC	Certification categories for seed of fodder, oil and fibre plants
Commission Decision 81/675/EEC	Non-reusable sealing systems
Commission Decision 80/755/EEC	Indelible printing on cereal seed packages
Commission Directive 75/502/EEC	Limiting marketing of *Poa pratensis* seed
Council Directive 71/161/EEC	Forestry reproductive material standards
Council Directive 70/458/EEC	On marketing of vegetable seed
Council Directive 70/457/EEC	The common catalogue of agricultural varieties
Council Directive 69/208/EEC	On marketing of seed of oil and fibre plants
Council Directive 68/193/EEC	Marketing of vine reproductive material
Council Directive 66/404/EEC	Marketing of forestry reproductive material
Council Directive 66/402/EEC	On marketing of cereal seed
Council Directive 66/401/EEC	On marketing of fodder plant seed
Council Directive 66/400/EEC	On marketing of beet seed

assurance for seeds (certain field crops, forage crops, vegetables, forest trees) and seed potatoes, including plant variety rights, national listing, common catalogues of varieties, and derogations. The principal items of legislation dealt with are listed in Table 4.2.

Standing Committee on Propagating Material of Ornamental Plants

Deals with regulations relating to marketing and quality assurance for propagating material of certain ornamental species. The principal items of legislation dealt with are listed in Table 4.3.

Standing Committee on Propagating Material and Plants of Fruit Genera and Species

Deals with regulations relating to marketing and quality assurance for propagating material and plants of certain species of fruits. The principal items of legislation dealt with are listed in Table 4.4.

The latter two committees usually meet on consecutive days, as it is usual for the same officials in the Member States to deal with both subject areas.

Voting

Voting procedures are set out in the Treaty of Rome, but have been amended in various ways. All proposals for decision by voting must be

Table 4.3. Principal legislation dealt with by the Standing Committee on Propagating Material of Ornamental Plants.

Commission Directive 1999/69/EC	Supervision of ornamental propagating material suppliers (repealing Commission Directive 93/63/EEC)
Commission Directive 1999/68/EC	On requirements for lists of ornamental varieties (repealing Commission Directive 93/78/EEC)
Commission Directive 1999/66/EC	Requirements for labelling of ornamental propagating material
Council Directive 98/56/EC	On marketing of propagating material of ornamental plants (repealing Council Directive 91/682/EEC)
Commission Directive 93/49/EEC	On health standards for ornamentals, as amended by Commission Directive 1999/67/EC

Table 4.4. Principal legislation dealt with by the Standing Committee on Propagating Material and Plants of Fruit Genera and Species.

Commission Directive 93/79/EEC	Requirements for lists of fruit plant varieties
Commission Directive 93/64/EEC	Supervision of fruit plant suppliers
Commission Directive 93/48/EEC	On health standards for fruit plants
Council Directive 92/34/EEC	On marketing of fruit plants

made by the Commission. In certain important areas of government, decisions by the Council are required to be made unanimously. However, in other areas and in the standing committees concerned with plant health, they are normally made by the system of weighted majority voting, each Member State having at present between 2 and 10 votes, approximately in proportion to the size of its population, as detailed below. In the plant health standing committees, Commission proposals are: (i) adopted by a qualified majority in the standing committee; (ii) adopted by a qualified majority in the Council; or (iii) amended by unanimity in the Council.

Where no decision has been reached after a period of, normally, 3 months following submission of a proposal to the Council, the Commission can adopt the proposal except where: (i) in the case of *filet* procedure, it has been rejected by a qualified majority; or (ii) in the case of *contre filet* procedure, it has been rejected by a simple majority.

The *filet* or *contre filet* procedure is specified for use on matters submitted for decision by Council according to the type of committee from which they come, or according to the particular clause of legislation from which they arise.

At the EU Summit (Heads of Government) meeting held at Nice, France, in December 2000, the numbers of votes per Member State were revised in preparation for enlargement of the EU by admission of applicant countries (mainly in eastern Europe) over the succeeding years (Table 4.5). Under the Treaty of Nice, potential votes were allocated to these countries, to be taken up on accession. In addition, the policy areas subject to qualified majority voting were extended. The Treaty of Accession (Athens, April 2003), after ratification by all existing Member States and the ten acceding countries, will admit the acceding countries as from 1 May 2004. Until 31 October 2004, the existing voting system will be extrapolated to include the new Member States. As from 1 November 2004, the revised voting system defined in the Treaty of Nice will apply (Table 4.5). A Member State may delegate its votes to another Member State, with or without instructions as to how the votes are to be used.

There are certain agreed provisions that affect the adoption of decisions by qualified majority voting if the vote is close. Also, in practice, the Commission is usually reluctant to take a vote on a proposal

Table 4.5. Number of Council votes per Member State and acceding countries, as agreed at Nice, 2000, and Athens, 2003.

	EU Member States, 2003			EU acceding countries, 2003	
	Before 1 Nov. 2004	After 1 Nov. 2004		After 1 May 2004	After 1 Nov. 2004
Austria	4	10	Cyprus	2	4
Belgium	5	12	Czech Republic	5	12
Denmark	3	7	Estonia	3	4
Finland	3	7	Hungary	5	12
France	10	29	Latvia	3	4
Germany	10	29	Lithuania	3	7
Greece	5	12	Malta	2	3
Ireland	3	7	Poland	8	27
Italy	10	29	Slovakia	3	7
Luxembourg	2	4	Slovenia	3	4
The Netherlands	5	13			
Portugal	5	12			
Spain	8	27			
Sweden	4	10			
UK	10	29			
Total	87			124	321
Qualified majority	62			88	232

if it is evident that a substantial number of Member States will vote against or abstain. In these circumstances, a renewed attempt is normally made to reach a compromise on which there is more nearly a consensus. Because of this desire to reach decisions that are as nearly unanimous as possible, it is often necessary to compromise on many points. Also, time for discussion is not unlimited. The result is therefore often less clear, concise and logical than it might be, or than is desirable, and it is thus not reasonable to expect the resulting legislative texts to be tightly worded or free from ambiguities. It is also possible for errors or changes of meaning to get in during the processes of translation, or even during preparation for publication in the *Official Journal*, as is required for all EC legislation.

The Plant Health Directive, 2000/29/EC

Council Directive 2000/29/EC (consolidating the old and much amended Directive 77/93/EEC) is the principal piece of legislation governing plant health regulations within the EU. Together with the implementing Commission directives (and others listed in Table 4.1), it governs the operation of the EU plant health regime. Its main provisions are shown in Table 4.6.

Table 4.6. Council Directive, 2000/29/EC: articles and annexes, with brief descriptive titles.

Articles
1	Scope. At Clause 4 is the requirement for each Member State to establish a single central plant health authority under the control of the national government
2	Definitions
3	Required bans on certain harmful organisms
4	Required bans on certain plants, plant products and other items
5	Items banned unless special requirements are met
6	Official export examinations; registration of producers and importers
7–9	Issue of Phytosanitary Certificates and reforwarding
10	Issue of plant passports
11–12	Action on material which does not conform to required standards
13	Imports from third countries; border inspection posts (BIP) facilities (see also Article 25)
14	Procedure for amendments to Annexes
15	Derogations
16	Emergency procedures; notification of the presence of listed or unlisted harmful organisms
17–19	Committee voting procedures
20	Procedure for amendments to the Directive; interaction with other legislation; special measures
21	Monitoring application of the Directive
22–24	Financial compensation provisions for losses and control measures
25	Community financial assistance for BIP facilities (ref. Article 13)
26	Reporting by the Commission to Council on financial assistance and compensation under Articles 13 and 22–24
27	Repeal of Council Directive 77/93/EEC and amendments
28–29	Timing and address

Annexes
I	Harmful organisms to be banned		
	A	In all Member States	
		i	Organisms not known to occur in the Community
		ii	Organisms known to occur in the Community
	B	In certain protected zones	
II	Harmful organisms to be banned if on certain plants or products		
	A	In all Member States	
		i	Organisms not known to occur in the Community
		ii	Organisms known to occur in the Community
	B	In certain protected zones	
III	Plants, plant products and other items to be banned		
	A	In all Member States	
	B	In certain protected zones	
IV	Conditions for movement of certain plants and other items		
	A	In all Member States	
		i	Originating outside the Community

Continued

Table 4.6. *Continued.*

	ii		Originating within the Community
	B		In certain protected zones
V			Imported plants, products and other items requiring health inspection in the third country or at the place of origin within the EU
	A		Originating within the Community
		i	Plants, plant products and other items requiring a plant passport
		ii	Plants, plant products and other items requiring a plant passport valid for the relevant protected zones
	B		Originating outside the Community
		i	Plants, plant products and other items requiring a plant passport
		ii	Plants, plant products and other items requiring a plant passport valid for the relevant protected zones
VI			Plants and plant products to which special arrangements may be applied
VII			Model certificates
	A		Model phytosanitary certificate
	B		Model reforwarding phytosanitary certificate
	C		Explanatory notes
VIII			Repealed Directive 77/93/EEC and its successive amendments
	A		Details of the repealed Directive and amendments, with exceptions noted
	B		Deadlines for transposition and/or implementation
IX			Table of correlation with Directive 77/93/EEC

Format

Like many other directives, 2000/29/EC consists of Articles (of which there are 29) and Annexes (of which there are nine). Much of the technical matter is contained in Annexes I–V. These annexes are divided into parts A and B, normally with parts A applying to all Member States and Parts B applying to certain protected zones. In Annexes I, II and IV, Parts A are further divided into Sections i and ii, with Sections i dealing with organisms or items from outside the Community and Sections II dealing with those from within the Community. Annex V differs from this pattern in that Parts A and B deal, respectively, with items from within and from outside the Community, while Sections i and ii, respectively, apply to all Member States and to certain protected zones. The articles, annexes and their main sub-divisions are shown in Table 4.6 with brief descriptive titles.

Some notable special features of the EU plant health regime are described below.

Transparency

In international parlance, 'transparency' in legislation and other matters means that the legislation and the underlying reasons for the legislation or action are clearly apparent and that there are no hidden conditions qualifying stated requirements or legal provisions. For example, in the EU plant health regime there is no general provision for the issue of import permits for the import of plants or plant products for the purposes of trade: all conditions relating to the control of such imports are as in the published legislation available to all and no further conditions may be imposed by the EU or the Member States. However, there are provisions allowing Member States to take emergency measures in cases where new serious risks to national or Community plant health suddenly arise, or to import, under special conditions, small quantities of normally prohibited items for the purposes of scientific research or trials. Transparency is a goal of all EC legislation, although its often convoluted nature sometimes makes interpretation difficult.

Bans and prohibitions

In common with bans on items of plant health risk contained within the plant health regulations of many sovereign states, the EU plant health regime prohibits the introduction to and spread within the EU of certain seriously harmful organisms (quarantine organisms), which are either not already present within the EU, or which are of limited distribution within the EU and are subject to control measures. These are the organisms listed in Annexes I and II, which correspond closely (but not exactly) to the EPPO A1 and A2 lists of quarantine organisms (see Appendix I Plant Health). The Directive Annexes differ from the EPPO lists in the way they are subdivided and in that a few organisms included in the EPPO lists are not mentioned in the Annexes. These discrepancies are mostly due to either the wider geographic coverage of EPPO, the slower administrative procedures needed to amend the Annexes or to lack of certain information desired before a decision is taken. Although all organisms are carefully considered before addition to either list, the fact that it is obligatory for Member States to comply with EC legislation tends to generate greater caution in amending the Annexes.

Some plants, plant organs that could be used as propagating material, certain fruits, wood, bark, soil or growing media, are considered to present so great a risk of introducing quarantine organisms that their import to the EU is wholly prohibited. This may depend on their geographical area of origin. In most cases, however, import to the EU is permitted provided certain conditions have been fulfilled. Indeed it is a particular feature of the EU legislation that all imports are permitted except for a very limited number of items that are restricted or prohibited. This contrasts sharply with the phytosanitary regulations of many nations, which prohibit or tightly control most items and require most imports to be made under licence or permit. The protected zones are always treated as special cases and carry their own conditions.

The prohibition of many harmful organisms depends on whether they are present on certain plants or plant products. It is considered that the risk of introduction and spread is usually greatly increased if these organisms are present on seeds or living plants, especially if they are planting material. The risk is normally lower if they are present on material that will be consumed or processed in some way, such as grain for consumption. Many harmful organisms inhabiting timber are easily transported with the timber trade and therefore are prohibited if associated with wood.

Plant passports

Certain plants and plant products are identified as associated with risks to plant health in the Community. For the movement and marketing of these plants and plant products within the Community, *including within the Member State of origin*, they must be accompanied by a plant passport giving relevant details and assurances as to the health of these items. Phytosanitary certificates are not required for this trade. Only those species and items considered to present a significant risk of spreading quarantine organisms are required to have plant passports, the aim of which is to give assurance of adequate plant health status and to permit the origin of the traded material to be traced in cases where faults are discovered. The proper issue of plant passports is the responsibility of the Member State in which the traded material has been produced or to which it is imported from a third country. Plant passports are issued by the responsible official body (normally the NPPO or an independent organization delegated for the purpose), or by producers or traders authorized to do so by the responsible official body. Inspections for issue of plant passports are normally required to be done at the place of production at an appropriate time in relation to the growth of the plants or production of products. Plants and other items subject to passporting which are imported into the Community from third countries must be issued with a plant passport after entry and after passing proper phytosanitary inspection. Replacement plant passports may also be issued where the status of the material or the composition of the consignment changes (for example, where passported consignments are split or amalgamated). All authorized issuers of plant passports are required to maintain adequate records to permit trace-back. The authorities, or the final recipients of passported items, are also required to retain the plant passports from traded items for a period of 1 year.

Plant passports consist of a label and an accompanying document (normally the invoice or consignment note), but may be a label alone, provided it carries all the required information. In the case of seed potatoes, for example, the official label required under the seed potatoes legislation also functions as the plant passport. The information required begins with the statement 'EC – plant passport' and then gives the codes of the Member State and responsible official body, the producer's registration number, and the serial or batch number of the product. Finally, there must be a statement of the botanical name, consignment quantity and (where appropriate), marking of validity for the relevant protected zone ('ZP' and name) and, in the case of replacement passports, the mark 'RP' and either the name of the third country whence the item was imported or the Member State of origin.

Some plants and plant products that are considered to present a smaller risk to plant health are exempt from the plant passport regime, but are subject to a system of lighter controls designed to enable trace-back. Such items include ware potatoes (for consumption or industrial use), most citrus fruit, and certain kinds of planting material, which are prepared ready for sale to the final consumer and whose production is clearly separate from that of other, passportable, material. However, producers and traders of such items do have to be registered.

Passported material is subject to a system of official inspections during marketing to check that it complies with required standards. However, such checks may not be made at national boundaries and must be random spot checks; not targeted or made in a regular pattern.

Registration of producers

Producers of, or traders in, plants, plant products or other items that require a plant passport for marketing within the Community are required to be registered by the appropriate responsible official body and approved by it as fulfilling the requisite conditions. These are designed to ensure that the producer has suitable facilities and has suitably qualified staff who know their responsibilities (including that of notifying the plant health authorities of the occurrence of any quarantine organisms), are able to carry out timely and effective inspections, and are able to recognize the relevant quarantine organisms. Registered producers and traders are required to keep records and documentation relating to the planting, production, trading or storage of relevant items on their premises, to cooperate with the responsible official bodies or their representatives, and to carry out phytosanitary inspections in an appropriate and timely manner according to guidelines provided. Small producers or traders supplying only the non-professional local market may be exempted from these regulations.

Phytosanitary certificates

Often known as 'phytos', these are issued only for consignments of plants and other relevant items that require a phytosanitary certificate and which are destined for export from the Community to third countries. They follow the model certificate in Annex VIII of Directive 2000/29/EC, which in turn follows the format set out in the IPPC (see Chapter 3). The certificate confirms that the requisite inspections and conditions specified by the country of destination are believed to have been fulfilled. Phytosanitary certificates are issued only by the responsible official body designated under the IPPC (normally the NPPO) or under its direct supervision, and their issue cannot be delegated to private producers or traders as with issue of plant passports.

Derogations

Derogations are agreed exceptions to the regulations of the Community. They can apply to articles or items in the annexes of any Directive, or other piece of phytosanitary legislation. Before coming into force, they must be discussed in the relevant standing committee and be approved by the proper voting procedure. For example, a derogation may be agreed to permit importation to the Community of otherwise prohibited material, and it will normally contain special provisions to which the material must conform before qualifying for importation.

Emergency procedures

Member States that discover the presence or spread of a listed plant quarantine organism within their territory are obliged to notify the European Commission. This action is compulsory, whether the quarantine organism has been found as an outbreak on a crop, intercepted on an import from a third country, or found on material received in trade with another Member State. Notification is also required where a quarantine organism has been found to have spread into an area in which it was formerly recognized as being absent. The European Commission has set up a database program for this notification and for submission and compilation of statistics, known as EUROPHYT.

It may happen that a Member State discovers the presence of an organism unlisted in the Community legislation but which it considers presents a potentially serious risk to agriculture or the environment. In such cases, the Member State may take such emergency action as it considers appropriate (under Article 16 of the Plant Health Directive), pending consideration of the situation by the Standing Committee on Plant Health. The Member State will be required to present a reasoned account of the situation and the reasons why particular action was taken, including a risk analysis in support of such action (see Chapter 11). The Standing Committee may then confirm the action taken, either with or without modification. If the action is not confirmed, the Member State must ensure that the action ceases forthwith.

Protected zones

These are areas within the EU where certain seriously damaging pests, which are present elsewhere within the EU and are not normally regarded as Community quarantine organisms, are not native or established, and in which certain extra phytosanitary measures against the relevant organisms must be taken. Member States wishing to have such areas recognized as protected zones (also known as *Zona Protecta*, or ZPs) must apply to the Commission and must follow a procedure laid down, which involves regular and systematic surveys to show the distribution or absence of the relevant organism, and provision of information about any official eradication measures being taken in cases where the relevant organism occurs in the area. There are certain time limits and other requirements that must be observed. ZPs may also be established in areas where certain crops are particularly vulnerable to certain harmful organisms, even though these are not native or established in the Community. ZPs may be all or part of a Member State, or several Member States, and movement of plants or other relevant items into or through such areas must observe the relevant extra phytosanitary measures laid down in the Annexes to Directive 2000/29/EC.

Third countries

These are all countries other than EU Member States. Certain of these countries may be recognized by the EU as having equivalence with the EU and can therefore be treated similarly to a Member State for the specific item and purpose concerned (for example, Switzerland is recognized as having equivalence for the marketing of seed potatoes).

Financial compensation

Articles 22–25 set out a framework of rules for financial assistance from the EU. This can be to bring facilities at BIPs up to requirements (a maximum contribution of 70%) or to compensate for costs and losses incurred in eradicating or controlling serious harmful organisms, whether listed or unlisted. In the latter case, the maximum compensation is up to 50% (25% for loss of earnings). It is also open to Member States or individuals to initiate claims under Community law for compensation where negligence or non-compliance with Community law can be shown.

Meetings of Chief Plant Protection Officers

These Heads of NPPOs from each Member State usually meet annually to discuss and resolve matters of common interest which it may have been difficult or inappropriate to resolve in the Standing Committee on Plant Health.

The European Phytosanitary Inspectorate

This service was established as part of the implementation of the Single Market to monitor the operation of the Single Market plant health regime and to carry out such other tasks as may be requested by the Standing Committee on Plant Health. In conjunction with the official services, it monitors the inspections made by the national plant protection organizations of the Member States, investigates reported outbreaks of harmful organisms, and fulfils a troubleshooting and information-gathering role, visiting third countries for this purpose if necessary. For example, EU inspectors may investigate allegations of trade in substandard material, applications for derogations made by Member States, or check the surveys done to substantiate protected zones. The Inspectorate is small, comprising not much more than one inspector from each Member State (the number of inspectors is usually less than this), a Chief Inspector and a small clerical staff. The Inspectorate is within the Food and Veterinary Office of the Health and Consumer Protection DG, with headquarters located in Dublin, Ireland.

The Vademecum

The Standing Committee on Plant Health has also been involved with the development of a *Vademecum*: a set of guidelines covering plant health operations within the EU. This is intended to assist the European Phytosanitary Inspectorate and national plant protection organizations in planning and carrying out various phytosanitary procedures, and also to promote the harmonization of such actions throughout the EU. The first sections of the Vademecum agreed upon concerned *Solanum* potatoes and potato pests. Sections covering plants for planting, fruit plants, ornamental plants, forestry plants, bonsai plants, strawberry plants, polyphagus pests, forestry pests and inspection of produce from third countries are now available or in preparation.

Operation of National Plant Protection Organizations   5

Introduction

Phytosanitary regulation is a comparatively recent area of activity for most national governments and it is still in its early development in some less developed nations. Although being a party to international conventions and agreements requires compliance with certain organizational points, this does not mean that all phytosanitary services must be organized in the same way. On the contrary, there are many different ways of satisfying the international obligations and national needs. In practice, the organization of NPPOs to deliver plant health services varies enormously, often demonstrating that even seemingly unsatisfactory arrangements can be made to work, given determination and goodwill on all sides.

International Obligations

All nation states are encouraged to become signatories to the two main international agreements described in Chapter 3, the IPPC and the WTO-SPS. Information on these organizations and the texts of the agreements may be found on the websites http://www.wto.org and www.ippc.int/cds_ippc/IPP/En/default.htm

In addition, countries are also encouraged to become members of their local RPPO. Membership of these bodies imposes certain organizational obligations. First, a single central government authority to operate procedures for the notification of regulations is required by the WTO-SPS Article 7, while the IPPC Article VIII requires the establishment of an official NPPO. It is often convenient for these to be combined. Both the WTO-SPS (Article 7) and the IPPC (Article VIII) also require the designation of an enquiry or contact point for the dissemination of information on plant pests and national phytosanitary regulations. For these, it is sensible to satisfy both requirements with a single government post or office, which normally would be a part of the administrative sector of the NPPO. Member States of the European Union additionally have a requirement under Council Directive 2000/29/EC, Article 1.4, to have a single and central authority controlled by the national government for the coordination of, and contacts on, plant health matters (see Chapter 4). This requirement can also usually be satisfied by the same arrangements made to comply with the WTO-SPS and the IPPC.

Influence of the Form of Government

The most important factor affecting the structure of NPPOs is, of course, the type of government operating in the nation state. This can vary from the close political control of one-party nations, often with major presidential powers, to more or less loose democratic federations of one sort or another, with considerable autonomy for the sub-national or regional units. The number of levels in national government is usually reflected in the organization of the NPPO, the simplest structures being in nations having strongly centralized government, while the most complicated are found in nations that are composed of numbers of federated units. In federal nations, the individual parts of the federation usually will each have their own plant protection organization that performs some, if not all, of the phytosanitary control in their appropriate areas, while international contacts, and often some of the more general regulatory tasks, are done by the federal government.

In nations composed of federated states, the international obligations for a single national phytosanitary contact or 'single central authority' can be met by the federal government. In nations where federal arrangements are not straightforward, this can be done by the agreed designation of one of the plant protection organizations as the lead organization, with authority for national representation. This should have a single, and preferably non-politically appointed, chief civil officer below the position of the responsible national government minister, who is normally the minister responsible for agricultural matters. It is now becoming increasingly common for these responsibilities to be combined with those for environmental issues.

Legislation

All phytosanitary action that is not voluntary must be taken under the authority of appropriate legislation. This is especially important if any enforcement is necessary. National legislation can be in a variety of forms, but usually it is enacted at two or more levels: primary legislation and secondary or subsidiary legislation. Primary legislation normally provides a general framework of salient points and principles, which enables more specific statutes to be made as secondary legislation. Such primary legislation, and usually also its amendments, normally require to be passed through the full law-making process and generally need to be approved by the legislative assemblies and head of state. This can be a very lengthy procedure and there are usually many possible pitfalls, including the priority given to the legislation by the government of the day, which may not be high. This can delay or eliminate draft legislation during its passage through the legislative process. Secondary legislation normally does not have to pass through the full legislative process and can often be promulgated as a ministerial order, regulation, or other type of statutory instrument after passing through an abbreviated and comparatively quick legal process. It is therefore much easier to make or amend, and because of this most detailed phytosanitary legislation tends to be of this type.

Phytosanitary legislation can also be made by local government. In federal nations this is done by the states or other political units constituting the federation. If these are large and powerful, this 'local' legislation can assume the importance and complexity of national legislation, whereas in non-federal nations such local legislation may be restricted to phytosanitary matters of only minor importance, or may not deal with such matters at all.

The maximum penalties prescribed and those actually imposed on conviction for contravention of phytosanitary legislation are often relatively light. There is a danger that governments may not regard contravention of phytosanitary legislation with the same gravity as other crime, and that the legal officer trying such cases may not appreciate the full implications of economic loss or possible damage to the environment that such offences can cause. Where the potential penalties for contravention of phytosanitary regulations are seen to be less than the prospective losses that might be incurred in compliance (for example, with an order prohibiting movement of potentially infected material), this can create severe difficulties for phytosanitary authorities. When drafting or amending phytosanitary legislation, therefore, it is advisable not to set maximum penalties at too low a level, especially as this can rapidly decline further with the effects of inflation if it is not frequently amended. For this reason also, it may be advisable to set penalties in terms that increase with inflation and not as fixed monetary sums. The aim should be to set penalties at a level commensurate with the potential damage that contraventions could cause.

Phytosanitary Systems

Phytosanitary regulation depends on three main areas of activity: (i) policy formulation and coordination; (ii) scientific and technical advice and related research; and (iii) inspection and enforcement. These three disciplinary sectors must be represented in any efficient NPPO, but within the organization can be combined or separated to varying degrees, depending on national circumstances. It is usually beneficial for these three areas to be under a unified command, but sometimes historical factors make this difficult to achieve. Nevertheless, it is frequently possible to make arrangements for operation, which create a similar effect. For example, a unified control of the budget for all three sectors, or provision of inspection or scientific services under contract, can often allow the head of the NPPO to exercise effective control. Where this is not done and one of the sectors remains under totally separate command from the others, the result can be confusion, particularly in areas such as prioritization of tasks and allocation of resources. This can be counterproductive, encouraging unnecessary administrative complications and unhelpful rivalry. It also will contravene the international obligations mentioned above. Each of the three sectors has firm links to each of the others and a strong unified or coordinated service should result from mutual support.

Policy and administration

As in most areas of government, well-considered and technically justified policy decisions are the key to efficient programmes and services. This administrative activity benefits by being part of core government in order to reflect current government policies and to ensure that the NPPO is regarded as an integral part of the government of the day, with consequent authority. Although there are examples where the delegation of phytosanitary policy to peripheral government agencies or government-sponsored bodies works well, such an arrangement may lay it open to undue or inappropriate sectoral influence, which usually detracts from its effectiveness and may encourage neglect by core government, both administratively and financially.

There are as many styles of policy making and organizational structure as there are governments, and different styles can be equally effective. However, to avoid impractical policies and waste of resources, policies need to be based on up-to-date and reliable information. This can be obtained by close cooperation with both the other sectors of the NPPO and by effective consultation with interested stakeholders (such as the relevant parts of the agricultural and forestry industries, traders, and appropriate environmental groups) before important decisions are taken or new initiatives started. Failure to do this risks antagonizing the sectors using the phytosanitary services and could result in the failure of otherwise sensible policies, as well as waste of resources. Proper consultation paves the way for informed but independent policy making by government and should result in fewer disputes between stakeholders and government at a later date. Care must be taken to avoid any special favouring of the interests of a particular sector, either in fact or in appearance. The NPPO must be, and be seen to be, impartial and fair in all its dealings.

Besides policy formulation, policy units often also undertake routine administrative tasks, such as international contacts and liaison with other parts of national government. Negotiations and discussions, both international and intranational with various interest groups, occupy much effort in the day-to-day operation of phytosanitary services. The lead in these is normally taken by the policy and administrative sector, but as scientific and practical problems or concepts are frequently considered, support by specialist scientific or inspection personnel is often essential. This is particularly so when major initiatives on surveys or eradication campaigns are being planned.

The essential administration of the service may be undertaken by a central group, perhaps attached to the central policy unit, or by the operations part of the service. It may include the administrative supervision of eradication campaigns (Chapter 7), the issue of permits and licences, the administration of healthy stock schemes (Chapter 8), making notification reports, compilation of statistics and maintenance of records. The administration often also acts as facilitator in making arrangements for provision of services for other parts of the service. For example, in arranging for translation of foreign regulations before they can be interpreted scientifically.

Scientific advice and research

The main tasks of the scientific part of a NPPO are: (i) to provide laboratory services for identification of harmful organisms in support of phytosanitary inspections and surveys; (ii) to provide advice on and interpretation of the scientific aspects of national and international legislation or other regulations; (iii) to carry out pest risk analyses; (iv) to provide scientific support, training and advice to other parts of the NPPO during negotiations, or when planning policy or executing initiatives such as surveys or eradication campaigns; and (v) to advise on the scientific aspects of certification schemes and issue of licences and permits. Ancillary but complementary to these tasks is the need to carry out related research to improve testing or identification procedures, to investigate pest biology where necessary to help in assessing threats and developing effective eradication measures, and to keep up to date with scientific advances in relevant disciplines.

The identification of plant pests (Chapter 10) is a highly skilled and specialized area of work which can involve the most technologically advanced procedures in biology. Plant pests belong to enormously diverse groups of organisms and many belong to groups that are taxonomically difficult to differentiate and classify. New species and strains are constantly being described and new techniques developed. Expertise in key taxonomic fields is therefore desirable in the scientific arm of an efficient NPPO. Scientists in the phytosanitary field need to be familiar with exotic pests that do not occur in their home nation, but which may arrive on imported material or by natural means, as well as with indigenous and established pests. It is therefore not possible for one, or even a few, scientists to be competent in the taxonomy and identification of all plant pests and harmful organisms, or to keep up to date with all new developments. This is a major difficulty for smaller countries with more limited scientific resources. A solution is to contract out the more specialized areas of pest identification to competent laboratories elsewhere or even in other countries. This is sometimes not politically popular, but with rapid modern communications it is often no slower than if the job had been done in the home country, and it has advantages in addition to securing accurate identification of difficult pest species. First, it is often cheaper than trying to maintain laboratory facilities and expertise within the NPPO or in the home country which may be needed only occasionally or for dealing with only small amounts of material. Second, it has the advantage of being seen to be independent and impartial. There are examples of NPPOs that have opted for such out-of-country arrangements, even though capable of doing the identifications within their service or elsewhere domestically.

Expertise in diagnosis of exotic pests is difficult for diagnosticians to acquire since, by their training and experience, most scientists will be more familiar with pests that occur in their own country. Training and foreign experience can provide a partial solution, and continuing experience in dealing with pests intercepted on imports will help to improve abilities in this task. It is also helpful if some forewarning of the likely occurrence of pests can be obtained. This can also be difficult because relevant scientific papers are not normally published for some considerable time after new pests appear (except in on-line journals). So, for advance warning of the spread of new pests, more reliance has to be placed on so-called 'grey literature': the trade and popular press of trading-partner countries. It is these weekly or monthly journals that are most likely to carry reports of pests that are becoming increasingly troublesome. However, considerable efforts in interpretation of such reports may be necessary and this can be time-consuming. Apart from possible problems with language, pest reports in such journals are seldom authoritative, and individual reports will often not be clear as to exactly what species is being referred to, where it has occurred, or even what kind of damage it has done. Nevertheless, taken together over a period of time, a pattern can be assembled from such reports, which may prove very helpful in providing early warning of hitherto unknown or unimportant pests that are increasing in their potential for damage and starting to move in trade or spread to new areas. It can also be difficult to obtain such journals, which often are not widely marketed outside the country of origin. In this respect the agricultural attachés or other representatives of the home country can be very helpful in obtaining, and even perusing, these journals.

PRA is a specialized and highly skilled discipline, which has greatly increased in importance in recent years (see Chapter 11). This is because international obligations and many national phytosanitary activities now need scientifically defensible PRAs to support any restrictions on trade. PRA lies within the scientific area of an NPPO because most of the analysis will deal with scientific aspects of the pest and its biology. It often has to be done at short notice and to a tight deadline, so it is necessary for the scientific personnel who may be called upon to do this work to be familiar with the appropriate formats and resources. Because the scientist doing the PRA must have a thorough knowledge of the pest and its relatives, a firm grasp of appropriate scientific methods, and a broad understanding of agricultural systems, it is usual for PRAs to be done by a senior scientist who is conversant with the pest group concerned.

There is little doubt that scientific research intended to improve phytosanitary abilities closely related to the operation of an efficient NPPO (such as in the fields of pest detection, identification and eradication or containment) should be an integral part of the scientific arm of the service. However, in countries with sufficient resources, there is often considerable debate as to whether government should fund research that is relevant to plant health but less closely related to the daily operations of the service. In countries where the national budget obviously cannot support much basic research, this debate may not be relevant. This can be a controversial debate, but it is important that basic research in fields related to plant health should not be neglected. Funding may come from either the public or the private sector, or both, but if private funding predominates, there should be some agreed arrangement for compulsory and enforceable contributions, so that the amount of the scientific budget for this purpose will be both known and adequate from year to year. Basic research cannot be stopped and restarted at short notice. There must be a continuing core programme, which not only adds to knowledge of the subjects but also retains and improves the expertise of key scientific personnel. Without this, expert personnel will move away to more secure and rewarding posts. The programme will decline or come to a halt, and the expertise will not be available when it is most needed.

Inspection and enforcement

This area of activity is usually covered by a specialized cadre of phytosanitary inspectors, forming a phytosanitary inspectorate. In some countries, especially in those with very limited resources, there is a temptation for the duties of phytosanitary inspector and agricultural advisor to be combined. However, this should be avoided wherever possible as it inevitably leads to conflicts of interest and detracts from the reputation and efficiency of both services. Phytosanitary inspectors need good basic scientific and agricultural qualifications, as well as training in the more specialized aspects of the work, such as the phytosanitary legislation, techniques for inspecting various plants, plant materials and other objects under very varied circumstances, and in the recognition of plant pests. They need to be familiar with current horticultural, agricultural and forestry practices and, as they are usually the public face of the NPPO, they need to be able to carry out their duties in a friendly but firm fashion.

It is normally convenient for such an inspectorate to be organized on an area basis, with individual inspectors responsible for certain administrative areas or border entry points. It may also be useful to form groups of inspectors with specialized expertise to operate nationally or across several administrative areas as and when the need arises. For example, teams specializing in seed potato inspection or in pest surveys may be composed of individuals from many different districts. Depending on the size of the country, there may need to be more senior inspectors responsible for organizing and coordinating inspection work over larger areas or in certain sectors of activity. Whatever organizational structure is adopted, the phytosanitary inspectorate should be responsible to the head of the NPPO and should have good and rapid means of communication, both within its own organization and with other parts of the service.

Provided a good and cooperative relationship is fostered between the phytosanitary inspectorate and the sectors of industry with which it comes into contact, much of the work will not require enforcement. In general, it is helpful to aim at operating with as little need for enforcement as possible. However, situations will inevitably arise in which necessary action will not be taken unless it is enforced. For example, this may be for the destruction of infected or at risk crops, or for the prevention of movement of goods constituting a real or potential plant health risk. All enforcement actions should be clearly taken under specific legislation and embodied in written Enforcement Notices or other documents. Likewise, the ending of action or lifting of restrictions should also be notified in writing. There should be close links with the customs and postal services, particularly for notification of the arrival of goods requiring phytosanitary attention at national points of entry and via international mail, and with the police for back-up, where necessary, in enforcement situations.

Communications

Good communications have always been essential to NPPOs. When such services were first being formed, communication was normally in writing and the telephone or telegram was used for more rapid dissemination of information. This formerly worked very well. Standing instructions to inspectors were also issued as hard copy, either in the form of a series of bulletins, or as a handbook, often in loose-leaf form. As more rapid or convenient means of communication became available, this was usually quickly adopted by phytosanitary services. In the 1980s, telex services became more widely available, but these were soon superseded by the telephone-based facsimile (fax) system, which was more user-friendly. In the 1990s systems of computerized communication developed with almost explosive rapidity, making available computerized networks, electronic mail (e-mail) and, more recently, dissemination of information via the Internet. The concurrent development of the mobile telephone and its linkage with e-mail and the Internet has now conferred facilities for communication undreamed of until the late 1990s. International communication by e-mail is now the norm, and notifications to international bodies of interceptions or the spread of pests are normally made by this means.

Currently, the usual aim is for all offices and laboratories of a NPPO to be linked by a computer network, with individual PCs or workstations for all but the most junior officers. Such networks are often nationally organized with secure access and sometimes are part of larger government networks. The transition from a system based on traditional, hard-copy methods to a computer network can be difficult and confusing. Programs for the operation of phytosanitary communications networks are often written and developed by national computer programmers and tailored specifically for the NPPO. These can work well, and at present there are few such programs available commercially. However, there is sometimes a tendency for locally written programs to become too complicated and sometimes they may be incompatible with other systems. Problems may also arise at a later date when, due to changes of IT personnel, the program has to be maintained by those who did not develop it. Detailed written records of program construction should therefore be made at the time such programs are developed.

Off-the-shelf packages for phytosanitary communications and information systems are now becoming available, but they need careful evaluation to ensure that they meet all the necessary requirements of the country concerned. The local RPPO will normally be able to provide advice. This is an extensive and complex subject and cannot be dealt with in detail here. For any country contemplating such an installation, one of the first actions should be the appointment of a competent systems analyst and IT specialist, as there are many pitfalls and unnecessary expense can easily be incurred in purchase of both hardware and software. Nevertheless, the cost of computing, particularly hardware, is decreasing and the advantages of an efficient system are so great that the system would probably save its cost in a relatively few years.

Computerized phytosanitary communications networks permit the rapid dissemination of instructions and the results of laboratory tests, the interrogation of databases, assembly of statistics, automatic printing of certificates, and all manner of other communications and acquisition of information. Inspectors in the field and other travelling officers should also have access to mobile telephones for immediate contact with headquarters and other parts of the service.

Written instructions for the NPPO as a whole, and phytosanitary inspectors in particular, are desirable to ensure uniformity of action throughout the service and to demonstrate that officers of the service are working to official and impartial rules. Desk instructions for the administrative sector are routine in most civil services, while standard operating procedures (SOPs) are also a normal part of biological scientific laboratory work (Chapter 10). Phytosanitary service instructions have to be revised and rewritten very frequently to reflect changes in international and domestic legislation and regulations, the changing status of pests, or the development of various services and campaigns. Historically, the most convenient way to produce printed copies of these instructions has been in loose-leaf format. However, the revision, production and distribution of instructions and their amendments is time-consuming and, where the telephone service is reliable, it is convenient to make instructions available on-line. Responsibility for production of phytosanitary instructions should be given to a particular officer or unit, assisted by *ad hoc* committees composed of relevant experts and experienced representatives of relevant parts of the service. The written instructions should also promote coordination and uniformity of action by setting out the division of responsibilities for each area of operation and for each part of the service. This is also encouraged by the use of standard forms and the practice of routinely recording the details of visits, inspections and other action in official notebooks or directly on to specially designed computer pro-formas.

Training

Training is a most important feature of NPPOs. For the same reasons that instructions have frequently to be revised, personnel need frequent training or retraining to keep up to date with new and changing circumstances. Training topics need to cover the whole range of phytosanitary activities, ranging from administrative functions to inspection methods, plant and pest recognition, horticultural production systems, and factory processes. Knowledge of contingency plans for action on outbreaks of serious pests and rehearsals of these are essential. The recognition of quarantine and other serious pests by the symptoms they cause and a knowledge of their biology and behaviour is basic to all phytosanitary work. In particular, it is necessary for diagnosticians and phytosanitary inspectors to receive training in the recognition of non-indigenous pests and their symptoms on both native and exotic host material. Plant variety recognition is important in all certification schemes for high-quality planting material (Fig. 5.1) and for this, annual or biennial retraining is desirable. Knowledge of factory processes is often needed. For example, control of potato or sugarbeet pests may require a knowledge of the processes for production of potato crisps, potato chips or sugar, including the systems for disposal of wastes. Training in methods of inspection and sampling can often be combined with other topics dealing with particular crops or commodities. For example, training in the use of mechanical spears for sampling seed and grain can be included in courses on seed certification or grain export. Systems of inspection for field crops can be included in courses on seed potato certification or on pre-export inspections.

As with the preparation of instructions, it is helpful if one officer is given responsibility for the organization of training, assisted by *ad hoc* committees of relevant and experienced personnel. This is a complicated task, including the identification of topics and suitable trainee candidates, the selection of suitable venues, the appointment of good trainers, and the prerequisite that districts or border inspection posts cannot be left without a minimum of staff while training takes place. For many courses, suitable crops and plots will have to be planted (Fig. 5.1) or other plant material for training purchased. This, together with the booking of external trainers and the fitting of the training periods into the pattern of work, often requires preparations to be started at least a year in advance of the prospective training course. The preparation of

Fig. 5.1. Plant health inspectors receiving training in the field on the recognition of cereal varieties, using small plots grown for the purpose. (Photo: courtesy of PHSI, Defra. Crown copyright.)

an annual training programme is therefore good practice and allows easy checking to ensure that all personnel receive adequate and appropriate training and that the necessary topics are covered regularly.

Except for those stationed at border inspection posts dealing with large volumes of plant material, phytosanitary inspectors spend much of their time working alone or at most in pairs. In spite of modern communications they may feel somewhat isolated. Training courses rectify this to a certain extent. However, in most countries an annual or biennial phytosanitary conference, which brings together phytosanitary inspectors and other parts of the NPPO, is of great benefit in generating good morale, with a sense of unity and *esprit de corps*. It also creates an opportunity for all ranks to meet each other and for lectures by authorities on relevant but peripheral topics, which might not otherwise merit inclusion on training courses attended by smaller groups of personnel.

Equipment and Facilities

The administrative arm of an NPPO requires normal office facilities and the scientific arm requires scientific laboratories for biological work. Where such laboratories dedicated to phytosanitary work are not established within the public service, this work is sometimes contracted out to university departments or other institutions. Scientific equipment is discussed in Chapter 10. Although such equipment may be more or less sophisticated according to the budget available, the scientific equipment and facilities required for phytosanitary work are not particularly specialized, except for quarantine containment for plants or pests (Chapter 12). The phytosanitary inspectorate also requires good quality equipment and appropriate facilities to produce high quality and reliable results, and this must not be overlooked.

Phytosanitary inspectors need to be mobile and able to travel immediately and quickly to deal with potential problems at places of plant production or points of import, so they must have ready access to dedicated and reliable motor transport. Much of the necessary equipment consists of protective clothing of one kind or another, including waterproof jackets, trousers and headgear, rubber boots, overalls, and safety helmets for use where heavy items are being lifted. Boots with reinforced toe-caps may also be necessary. Disposable overalls and overshoes are frequently used, and appropriate disinfectant for use on footwear and utensils may need to be available. A phenolic-based disinfectant is usually effective against a wide range of bacteria and fungi, while tri-sodium orthophosphate is effective against virus pests (see Chapter 12). A good-quality folding knife and a hand lens should always be carried. Equipment for sampling and lifting plants will include garden forks and spades, trowels, knives, pruning secateurs, polypropylene bags of various sizes together with fasteners, marker pens, labels and reliable torches (flashlights). Soil sampling is also an important and frequent task for which suitable augurs, cheese scoops or trowels are needed. More specialized equipment is, of course, necessary for many tasks such as sampling seed and grain, for which mechanical or manual sampling spears are needed, and riffles for mixing and subdividing into official samples. Surveys may require special equipment for sampling, such as boats for sampling aquatic vegetation or respirators for conditions where dust or spray droplets may be a hazard. All but very minor fumigation operations require special equipment and so are more conveniently done by specialist contractors (Chapter 12).

Border inspection posts

Facilities needed for phytosanitary inspection at BIPs will vary according to the volume and type of material handled. Although in many cases, where the pest risk can be managed acceptably, inspection of consignments can be deferred to the point of destination, some inspection at BIPs will be required. The basic requirements are shelter from inclement weather (particularly rain, snow, wind and excessive cold or heat), adequate lighting or shading, and a bench or portable table on which to examine material. More specialized equipment will be needed at BIPs dealing with large volumes of material. In such cases an inspection room (Fig. 5.2) and an office cum field laboratory may be required. Equipment for handling material in large quantities, such as cranes and forklift trucks, will usually be available at ports, railway goods yards and airports, but may sometimes have to be provided by the phytosanitary service at BIPs dealing with road vehicle traffic. For this traffic it is convenient to have a platform of height approximating to the floor level of transport vehicles and permitting easy access at consignment level to both sides and rear of the load space. There should be enough adjacent space to unload vehicles and containers and stack a consignment awaiting inspection while also allowing adequate working space between the vehicle and the unloaded consignment. Other safety and health considerations may also need to be taken into consideration. According to the type of goods to be inspected, there may be a need for other facilities such as controlled temperature stores, for example, if perishable material might need to be held for more than a few hours, and machinery for emptying sacks and boxes, conveying material past an inspection point and repackaging it. Laboratory equipment at BIPs need not be elaborate, especially where reliable postal services are available for rapid transmission of samples to a specialist diagnostic laboratory. Where this is not possible, the main need is for adequate stereo- and compound microscopes with appropriate lighting,

Fig. 5.2. A typical inspection room at a border inspection post. Note adequate lighting, inspection tables, scales and adjacent office. (Photo: author.)

instruments for specimen preparation, an incubator and dishes for incubating (but not culturing) material, and reference works for pest identification. Research-standard microscopes are not required. Within the European Union the minimum facilities required at BIPs are specified in Commission Directive 98/22/EC.

In some countries the NPPO inspectors wear a distinctive uniform, particularly at BIPs. This has the advantages of giving the service a smart appearance, making inspectors easily identifiable to the general public, and giving them a recognizable official status. However, it will increase costs and also tends to equate the service's public image with that of the customs and the police, which may be somewhat counter-productive where a more low-key and personal relationship with service clients is being fostered.

Health and Safety

In most countries there will be legal requirements and regulations concerning health and safety, which must be complied with. There are no special health and safety problems peculiar to the administrative or scientific operations of a phytosanitary service. In these areas, health and safety considerations will be very similar to those connected with the operation of offices and biological laboratories for other purposes. However, there are some particular hazards to which the officers of the phytosanitary inspectorate may be exposed. In the field, or in protected crop environments, the main hazard for inspectors will usually be toxic pesticides that have been applied shortly before visits. To guard against this, inspectors should make a point of ascertaining the pesticide regime and what pesticides may have been applied before they enter crop areas or handle treated seed. They should then take appropriate precautions, if necessary, with advice from the scientific arm of the Service. The very varied nature of the situations in which inspectors may have to work makes it difficult to specify the hazards that may be encountered, but potentially dangerous situations should be avoided wherever possible. There should be a general awareness of possible risks, especially when working in areas near large, heavy or mobile machinery, vehicles (including fork-lift or dumper trucks), cranes, moving ropes and cables, or where loads are being lifted. These situations often occur on ships, in dockside areas or in factories. Where ladders are used, these must be in good repair and firmly placed or fixed. If it is necessary to ascend to considerable heights, the use of safety harness should be considered. Dust and mists or aerosols containing harmful chemicals or organisms present another health hazard, which may need to be avoided or mitigated by use of masks or respirators. Dust is particularly dangerous in situations where it may contain toxic pesticides or other chemicals, disease agents or allergy-inducing spores. Hazards may arise from entering closed vessels, storage bins or low-temperature facilities, and at farms and factories there may be dangers from effluent drainage pits. Each situation must be assessed and, where necessary, appropriate and reasonable precautions taken to minimize risks to health and safety.

Costs and Charges

Many governments provide phytosanitary services free of charge as part of the public service. This particularly applies to economies based largely on agriculture, especially where there is a desire to encourage and support exports. However, with the increase in trade in plants and plant products and in the ability of the producer to pay at least a part of the costs, there has been a trend in the more developed nations towards a fee-based service in order to stem the escalation of public expenditure and to arrange that the costs are borne by those who benefit from the service. This also helps to avoid allegations of unfair competition and inequality between trading partners with differing levels of support for their phytosanitary services. The basis for charging varies from country to country, often being based on the quantity of material inspected or the cost of a unit of phytosanitary inspection time. Sometimes a flat-rate fee is levied, which entitles the payer to a certain range or quantity of services. Even where charges are made, these may be below 100% recoupment level by means of government subsidy, or some sectors of the service may be exempt from charge. For example, laboratory tests for pests, which are subject to statutory controls, may be provided free, on the grounds that this is in the wider interest of the agricultural sector or the public as a whole.

Normally, no charges are made when plants or crops are required to be destroyed or treated as part of an eradication campaign. On the contrary, phytosanitary authorities sometimes pay compensation in these circumstances. However, it is possible that this may generate difficulties for the phytosanitary authorities (see Chapter 7).

Phytosanitary Arrangements in the UK

As an example of a nation where federal arrangements are not simple, it may be helpful to look briefly at the organization of the NPPO (phytosanitary service) of the UK, as operating in the year 2002. The full national title is the United Kingdom of Great Britain and Northern Ireland. Great Britain comprises England, Wales and Scotland, including the Orkney and Shetland Islands. The Channel Islands and the Isle of Man are self-governing British Crown Dependencies that are not part of the UK. The geographical term, British Isles, refers to the archipelago off the north-west coast of continental Europe, which includes the main island of Great Britain and the island of Ireland together with their subsidiary islands, including the Orkneys, Shetlands, the Isle of Man, and the Channel Islands.

The constituent territories vary in their degree of political autonomy (this is in the process of changing, as exemplified by the new Assemblies for Scotland and for Wales). There are separate phytosanitary services for: (i) England, which by agreement also covers Wales; (ii) Scotland; (iii) Northern Ireland; (iv) the individual Channel Islands of Jersey and Guernsey; and (v) the Isle of Man. In Wales the phytosanitary service of the UK Department of Environment, Food and Rural Affairs (Defra) operates under the authority of the National Assembly for Wales Department of Agriculture, Industry, Economic Development and Training. In the other territories, the phytosanitary service is a part of the local department responsible for agriculture. In Northern Ireland the scientific input to the phytosanitary service comes from the Queen's University of Belfast and university staff devote part of their time to this business. In England and Wales and in Scotland scientific services are provided respectively (under Memoranda of Understanding) by the Central Science Laboratory (an Executive Agency of Defra) and by the Scottish Agricultural Science Agency, an Agency of the Scottish Executive. For Great Britain, phytosanitary matters relating to forestry and forest products are dealt with by the Forestry Commission, with headquarters in Edinburgh. The Forestry Commission reports to the Secretary of State for Environment, Food and Rural Affairs for forestry in England, to the appropriate ministers of the National Assembly for Wales, and for Scotland to the Scottish Executive Environment and Rural Affairs Department. In Northern Ireland responsibility for forestry matters rests with the Forest Service, an Executive Agency within the Department of Agriculture and Rural Development for Northern Ireland.

For most purposes, common phytosanitary legislation covers the territories of Great Britain (England, Scotland and Wales) while separate but similar legislation operates in each of the other territories, although in the Channel Islands and the Isle of Man the legislation may reflect the much more limited scope of the agricultural economies. There are some minor statutory instruments that operate only in Scotland, or only in England and Wales. Defra handles international phytosanitary affairs and the Defra Plant Health Division contains the international contact points for formal phytosanitary contacts on behalf of the UK. The Channel Islands and the Isle of Man (which are not as yet fully part of the European Union) have direct international contacts with a certain amount of support from the UK. Formulation and coordination of UK phytosanitary policy and relevant initiatives are organized through joint standing committees and *ad hoc* meetings.

It can be seen, therefore, that the structure of the UK NPPO is very fragmented, not simple, and far from ideal. This structure has come about mainly through historical events that had little to do with agricultural interests. Nevertheless, it works reasonably efficiently, largely due to the mutual goodwill shown by all parts of the services, and is a good example of a less than ideal organizational structure working satisfactorily, as noted at the beginning of this chapter.

Imports and Exports   6

The International Framework

Countries wishing to export goods including plants, plant products, or items that could present a risk to plant health, have to satisfy the phytosanitary requirements and regulations of the importing country before it will accept such goods. The onus is therefore on the exporting country to ensure that this is done. This principle is incorporated in the wording of the Model Phytosanitary Certificate set out in the Annex to the IPPC (Anon., 1999a, and see Chapter 3), which states:

> This is to certify that the plants, plant products or other regulated articles described herein have been inspected and/or tested according to appropriate official procedures and are considered to be free from the quarantine pests specified by the importing contracting party and to conform with the current phytosanitary requirements of the importing contracting party, including those for regulated non-quarantine pests.

Such a phytosanitary certificate (often referred to as the 'PC' or 'Phyto') as the original document or, in certain circumstances, as a certified copy, must accompany each consignment of goods subject to phytosanitary regulation, except in circumstances where trading partners have specifically agreed to bypass or replace this requirement, as is the case with trade between Member States of the European Union (Chapter 4). Phytosanitary certification in electronic form is also acceptable, so long as the intent of such certification under the IPPC is achieved and with certain other provisos, as detailed in ISPM No. 12 (Anon., 2001b).

Precautions by the importing country

In most countries with a long history of international trade, phytosanitary regulations covering the conditions for the import of goods that could present a phytosanitary risk were introduced gradually over many years in response to incidents or perceived risks from plant pests. This is now referred to as pest risk 'management'. These regulations have usually evolved through many modifications, often to combat specific threats. More recently, and particularly for countries that are members of the World Trade Organization (WTO) and contracting parties to the IPPC, they have evolved to comply with the international initiatives for harmonization and scientific justification described in Chapter 3. In countries that have entered the international market in goods subject to phytosanitary controls more recently, phytosanitary regulations may have been adopted more or less from international or regional models. In such circumstances care will be needed to ensure that the phytosanitary regulations are appropriate to their particular circumstances, can be technically justified, and otherwise conform to the requirements of the WTO Agreement on the Application of Sanitary and Phytosanitary Measures (WTO-SPS) and the IPPC. This process can present serious problems to small or developing countries, in which there may be a lack of resources or scientific expertise appropriate for this purpose. Nevertheless, technical assistance in this field should be available through the local RPPO, from a neighbouring country under the initiatives of the WTO-SPS, or from the FAO.

Pest Classification and Pest Lists

For potential new imports of plants or plant products, the first step is for the importing country to make a PRA (Chapter 11) of the commodity proposed to be imported. The ISPM Nos 2 and 11 (Anon., 1996b, 2001a) provide guidance. One of the results of this process will be the identification of a range of pests to which the commodity may be a host, or with which it may be associated. However, there may be a large number of such pests to be considered. Inevitably, some pests will be assessed as more serious than others, and so some kind of classification for pests is helpful in determining appropriate responses in terms of exclusion, management or control, and eradication if appropriate. For phytosanitary purposes, plant pests are usually grouped into the categories of quarantine pests, regulated non-quarantine pests and unlisted pests.

Quarantine pests

Quarantine pests have been defined (ISPM No. 5; Anon., 2002a) as 'pests of potential economic importance to the area endangered thereby and not yet present there, or present but not widely distributed and being officially controlled'. Therefore, quarantine pests are those that are considered to be potentially very serious for the country concerned. Although generally absent, they may already be present within the country but, so long as they are not widespread and government action is being taken to control them, they can still be considered as quarantine pests and phytosanitary measures to prevent their import can be justified. The PRA for a quarantine pest will therefore highlight its potential for future damage or for further spread rather than its current status. Detection of a quarantine pest in the country or on an imported consignment will normally necessitate some kind of action for its control, elimination or eradication.

Phytosanitary regulations often divide quarantine pests into various categories. For example, pests may be listed as regulated in any circumstances, or only when in association with certain hosts or plant products. They may also be divided into: (a) those that do not occur in the country or region concerned and (b) those that do occur, but are not widespread and are subject to official control. RPPOs also may operate regional lists of quarantine pests divided into these categories. In the European and Mediterranean area, for example, EPPO maintains two lists: the A1 List, corresponding to (a), and the A2 List, corresponding to (b). EPPO has also published detailed datasheets on pests of recognized quarantine status for the Euro-Mediterranean region (Smith *et al.*, 1996).

Regulated non-quarantine pests

Some pests that do not fall into the category of quarantine pest, perhaps because they are already widespread, nevertheless may cause substantial economic loss. This may be not only by direct effects on yield or quality of produce, but through increasing the costs of pest control or crop operations, or by necessitating replanting earlier than in the absence of the pest. Losses may also be caused through the need to grow crop varieties that are lower yielding, or that are otherwise less desirable, but that are resistant to the pest in question. Where governments consider regulation to be appropriate, and where within the country they are subject to official control on planting material (including seeds), such pests may be classed as regulated non-quarantine pests. ISPM No. 5 (Anon., 2002a) defines a regulated non-quarantine pest as one 'whose presence in plants for planting affects the intended use of those plants with an economically unacceptable impact and which is therefore regulated within the territory of the importing contracting party'. ISPM No. 16 (Anon., 2002c) outlines the concept of regulated non-quarantine pests and its application. Non-quarantine pests that are officially regulated through tolerances in a domestic certification scheme for planting material (Chapter 8) will therefore usually fall into this category, provided the scheme is obligatory for the material in question. Non-quarantine pests that are officially regulated by the importing country on planting material in some other manner may also fall into this category. In these cases the same or equivalent health standards, or other phytosanitary measures, may be applied to such pests on imported planting material of the same species.

Unlisted pests

Not all organisms harmful to plants and plant products are as yet recognized as such, and many pests have not yet been taxonomically classified and named. Many are not recognized as pests in their native haunts due to a range of factors such as competition or predation. Some organisms that formerly were not particularly harmful to plants may become so through genetic mutation or because of a change in agricultural practices. For example, changes in the crop species or varieties cultivated, or the acquisition by a pest of resistance to a pesticide, may result in some pests which were formerly of minor importance meriting quarantine status. The status of pests is thus continually changing. Also, therefore, assessments of pests can never be final or complete, and must be kept continually under review with corresponding amendment of national phytosanitary regulations. Pests which do not feature in phytosanitary regulations as quarantine pests or as regulated non-quarantine pests, are referred to by Cannon *et al.* (1999) as 'unlisted pests'.

Unlisted pests, which were formerly unrecognized or even undescribed, frequently come to notice when they start to spread in international trade. Some such pests may not be considered appropriate for listing as quarantine pests at the time of their first assessment by PRA, but may become so later as a result of accumulated information or changing circumstances. There will also be a time lag between the recognition of a new quarantine pest and its addition to quarantine pest lists.

Review and modification of pest lists

It is imperative that lists of quarantine pests and regulated non-quarantine pests are kept under review and promptly modified as circumstances change. Pests may have declined in importance to such an extent that official action is no longer considered justified, and therefore they should be removed from the relevant regulated pest list. This could be, for example, because of changes in agricultural practices, the development of new control systems, or simply because they have become so widespread that they are an established feature of the production of the host crop. In some cases, former quarantine pests may merit control as regulated non-quarantine pests, but this must be technically justified.

Quality pests

A quality pest has been defined (Anon., 1996a) as 'a non-quarantine pest for an importing country whose presence in a consignment of plants or plant products has economic importance in so far as it affects the grade, marketability or ultimate use of the consignment, and which may be subject to control under relevant quality regulations'. These are usually common pests, which may be difficult or impractical to control by administrative means but which may affect the quality or yield of crops and produce when attacks are severe, and for which there may be tolerances in quality regulations that do not qualify as phytosanitary measures. Official control under phytosanitary regulations, therefore, is not appropriate.

General surveillance, pest lists and records

Countries that are contracting parties to the IPPC are obliged (under Article VII.2i) to 'establish and update lists of regulated pests'. These must be made available to the IPPC Secretariat, RPPOs of which the countries are members, and to other contracting parties on request. The ISPM No. 18 (Anon., 2003b) provides guidelines for the management of these lists. For assembling such lists, and for the proper administration of plant health matters, NPPOs need to be aware of the plant pest situation in their areas of responsibility, and for this they need to gather information on pest occurrence and status from all available sources and to maintain a national system of pest records (ISPM No. 6; Anon., 1997a). There are many sources of information on pests, both within a country and elsewhere, and these vary greatly in reliability, so information needs to be evaluated carefully. For example, lists of plant pests and their hosts have been published for many countries independently from their phytosanitary regulations. These lists draw together records from many different sources, but may not include information permitting a judgement on their reliability. In this case they must be treated with caution by NPPOs when assembling official national pest records. The basic information needed for a reliable pest record is set out in ISPM No. 8 (Anon., 1998a), which also details the many possible sources of information, recognizes their variation in reliability and provides guidelines for accurately describing the status of a pest in an area.

Many harmful organisms are difficult to detect and are often overlooked, perhaps because local host plants are tolerant to infection or attack and do not show overt symptoms, or because the organism is difficult to identify, or it may simply be because no search has been made. The absence of a record, therefore, may not be a reliable indication that a pest does not occur in the area concerned, unless a specific survey for its detection has been done. Conversely, a country may have records of pests that have been intercepted and identified on imports, but which do not occur naturally in the country concerned. This may not always be made clear in unofficial records and can lead to misunderstanding of the status of a pest. ISPM No. 8 gives phrases appropriate for describing the various circumstances in which a pest can be declared absent and lists references for terminology and taxonomy.

Import Regulations

General considerations

It is usually not possible to identify all the plant pest risks that potentially threaten a country. The phytosanitary risk presented by imports normally increases with the distance from their place of origin and with the degree to which the climate in the area of origin corresponds with that in the importing country. Imports from nearby areas are less likely to carry important alien pests because pests from such areas will normally have been received through trade in the past. If capable of establishing, they will probably have already established, while those not capable of establishing present little risk. Various types of goods also vary in phytosanitary risk. Propagating material and plants for planting are living material and are intended to be planted and multiplied, often in an open environment near other vegetation that may contain conspecific or related plants. Such material presents a high phytosanitary risk as a pathway for alien pests because it can support most types of pest, even those parasites confined to living tissue, and when planted will often be near other receptive or susceptible plants and be able to release pest progeny or propagules for dispersal by air currents, rain splash or by their own locomotion. At the other end of the risk spectrum are processed plant products, which do not contain living material or which have been subjected to processes likely to kill any harmful organisms they might have contained (e.g. tea and coffee). Between these extremes lies the vast trade in food and forage items, which will usually be consumed before pests can develop or spread to growing hosts.

Phytosanitary measures can be formulated to diminish (but not necessarily to eliminate) most phytosanitary risks. Generally, the simplest and most effective measure to exclude a particular pest is to prohibit its import and the import of its hosts or substrates. There may be a lack of information on certain species of plants or pests, but enough information to justify prohibiting the importation of plants belonging to certain plant groups or originating from certain geographical areas. Geographically isolated countries, such as islands, which tend to be particularly vulnerable to damage by alien pests, may be more justified in taking stringent measures than countries that are part of a large land mass and which already have a large range of pests. However, prohibition obviously prevents trade in such commodities and usually can be justified only for the most serious pests or those that are the most difficult to control by other means. With some pests (including most with an air-borne dispersal stage), in certain situations, it would be impractical or uneconomic to operate effective exclusion measures (for example, if they are already present in neighbouring countries). In practice, and to comply with IPPC and WTO-SPS obligations, measures can generally be formulated in such a way as to minimize adverse effects on trade while providing the minimum measures necessary to give a high chance of achieving the desired objective of reducing the risk to an acceptable level. Thus, it is usually possible to devise measures that allow trade to continue while still providing adequate protection against the pest or pests concerned.

A harmonized format for phytosanitary import regulations would greatly facilitate the retrieval of information, especially by foreign trade partners, and would help interpretation and to minimize misunderstandings. As yet, no such international standard exists, but it is to be hoped that one will be formulated before long. Standard headings could include some of the examples listed in the section on 'Information on foreign countries' phytosanitary import regulations' below.

Specific import measures

Phytosanitary regulations relating to imports usually begin with a list of quarantine organisms whose import is prohibited. This will usually correspond more or less to the local RPPO lists (a) and (b) mentioned above. As a minimum, the regulations will specify that imported consignments must be free from such organisms. In addition, for their major hosts and substrates, a phytosanitary certificate will usually be required, through which the NPPO of the country of export certifies that the consignment has been inspected and/or tested and found free from the quarantine pests specified by the importing country. However, in order successfully to exclude the specified pests, it may also be necessary to regulate, or even prohibit, the import of their hosts or substrates. Special measures may be required if the organisms can be present on imported consignments in a state or stage in which they are impossible or difficult to detect. For example, viruses of fruit trees that show symptoms only on the leaves or fruit will not be visually detectable if the trees are imported during the dormant season. On the other hand, importation during the dormant season, when leaves are absent, will guard against pests that affect only the leaves. In most cases, complete prohibition is not necessary as many less-restrictive measures can be applied. The more usual types of regulatory measures that are applied to plants, plant products or other items to reduce or eliminate a phytosanitary risk are:

1. Material required to be free of relevant symptoms.
2. Import only from declared pest free areas (ISPM No. 4; Anon., 1996d).
3. Importation restricted to certain times of the year, during which symptoms would be visible or susceptible organs absent.
4. Import limited to plants from crops that have been officially inspected or tested and certified free from the pest.
5. Material required to have been treated (e.g. with heat, cold, fumigation or pesticide) during growth or as part of preparation for export.
6. Material required to have been trimmed to eliminate the parts or organs likely to harbour particular pests (e.g. debarking of timber).
7. Plants required to have been propagated from parent or ancestral material that has been appropriately tested and maintained.
8. Material required to have been grown on land tested and found free from relevant pests, or which has not carried relevant hosts for a certain length of time.
9. Items required to be free from soil or organic debris.
10. A representative sample (e.g. of seeds) required to have been tested and found free from relevant pests.
11. Plants required to have been cultivated and not collected from the wild.

All of these measures will be applied prior to export and should be confirmed by the NPPO through the issue of a phytosanitary certificate. Other measures may be required either during transport or on arrival.

These include:

1. In-transit treatments, such as with heat, cold, fumigation or pesticide.
2. Retaining plants in post-importation isolation (quarantine), pending official inspection, testing or treatment by the authorities of the importing country.

In-transit treatments fall between the responsibilities of the NPPOs of both the exporting and importing countries, as at the time they are being performed they fall outside the jurisdiction of both services. Therefore they cannot be covered by the normal phytosanitary certificate. To be effective they require the close cooperation of all trade sectors involved, including in particular the carrier of the consignment (e.g. the ship operator and staff), and close monitoring of treatment records by the NPPO of the importing country on arrival. Quarantine procedures required on arrival or post-entry are wholly the responsibility of the NPPO of the importing country.

Import prohibitions

As mentioned above under the section concerned with general considerations, it may not be possible, for a variety of reasons, to devise requirements that permit the safe importation of certain commodities. Where the PRA indicates that prohibition is justified and is the only measure that provides an appropriate level of protection, then such a prohibition should be specified in the import regulations to ensure that traders are aware of the prohibition. The prohibition should be clearly defined as to the commodity concerned and as to whether the prohibition applies generally or just to a commodity of a particular origin.

Soil and growing media

One general item that is almost universally prohibited is soil, which is often traded as a constituent of plant growing media or in association with living plants. Soil contains an enormous microbial and microfaunal population, only a fraction of which is known and which varies with soil type and from place to place. Soil is very difficult to sterilize in large quantities, so when used in growing media it should normally be regarded as non-sterile, and its risk will vary with that of its place of origin. Many plant pests inhabit soil during all or part of their life cycle and often produce durable resting stages in the soil which may be difficult to detect. Normally, living plants moving in international trade are required to be virtually soil-free or to be in soil-less media. Planting material such as tubers, bulbs and corms are usually required to be soil free or to carry the minimum amount of soil practicable. Requirements covering plant produce for consumption can be less stringent because the risk of imported soil finding its way on to agricultural land is much less. Soil is often found as a contaminant or substrate in goods where it may escape attention. For example, bonsai or penjing ornamental miniature trees in decorative containers often carry soil as part of the substrate. Soil is also often overlooked, even when present in substantial quantities, when adhering to imported used vehicles, agricultural machinery, or military equipment that has been used in the field.

Growing media and composts often contain peat, tree bark, coir dust or other by-products of plant product processing, each of which must be assessed for its phytosanitary risk. Some materials, such as peat taken from below cultivation depth, present very little risk and may even be regarded as not requiring phytosanitary regulation. Others, such as bark, may present more or less risk depending on the tree species, place of origin and what treatment it has received.

Packing materials and wood derivatives

Packing materials or dunnage composed of untreated plant material (such as the baulks of rough timber frequently used to pack cargo in the holds of ships) can present a serious risk and may warrant a requirement for the material to have been treated. Packaging composed of soft woods, wood shavings, bark products, sawdust and straw are other examples of this. ISPM No. 15 (Anon., 2003a) provides guidance on this topic.

Potatoes

Tubers of the potato (*Solanum tuberosum*) are also usually subject to stringent restrictions or prohibition, owing to the importance of the potato in many agricultural systems and the ability of many potato pests to exist in a viable latent state in potato tubers. The main aim in this respect is to prevent tubers intended for consumption or industrial use (known as 'ware' potatoes) being used as planting material and so, where importation is permitted, restrictions often require their treatment with chemical sprouting suppressants to deter this. Similar restrictions could be applied to other species where organs traded as produce could be used for propagation and the crop is locally important. Trade in potato tubers intended for planting (so-called 'seed' potatoes) is almost entirely of material produced under recognized official certification schemes designed to produce healthy material, as in **4** and **7** above (see Chapter 8).

Plants in tissue culture and micropropagation

The form in which plants are traded can greatly affect their potential plant health risk. Plants traded in the form of micropropagated plantlets or as other forms of tissue culture have several advantages. Plant tissue culture is a highly specialized form of plant propagation, performed *in vitro*, which was first used widely in the 1980s and is described in Chapter 8. It has many advantages for international trade and for propagation of certain plant species, including many that are difficult to propagate by other means. Tissue cultures are small, easy to transport and are often traded between countries. Because the material is grown on sterile, artificial culture media within glass or plastic containers, it is protected from incoming infection. So, provided the starting material is healthy, plants in tissue culture carry very little plant health risk, and material that might otherwise be prohibited could be permitted entry. Micropropagation is a form of tissue culture in which a relatively large shoot (about 5 mm long) is used as the starting material. However, plants propagated by tissue culture cannot be assumed automatically to carry no phytosanitary risk merely on account of their means of propagation.

Import permits for commodities in trade

Many countries use these documents to specify the quantity and the conditions under which all or certain specified plants and plant products may be imported. Where this is done, even routine imports of these commodities are not allowed without the necessary permit. Other countries (notably those belonging to the European Union) do not employ import permits because they are regarded as an impediment to trade and are not 'transparent' if their detailed conditions are not published in advance. However, for trade development, import permits can be used to great effect and in conformity with international obligations. Where there is currently no trade, or a particular trade is prohibited because of lack of information and experience, it is often difficult to determine pest risks and to select import measures that will provide the necessary confidence to the various trading partners involved. Import permits allow procedures to be established and tested under controlled conditions. If effective, then the measures can be incorporated in the normal phytosanitary import regulations without the necessity for specific permits.

Licences for exceptional imports

Regulations covering imports should normally have the possibility of exceptions being made for special or exceptional circumstances, subject to the discretion of the responsible government. This facility is often needed where prohibited material is required for scientific research, or when a breakdown in the normal supply chain has caused an acute need for a certain product, such as a foodstuff or crop planting material. It is normally operated by means of a licensing system. A licence is a document, issued under the relevant exceptions clause of phytosanitary regulations, which permits importation or certain other action which would otherwise be prohibited. The licence should specify in detail what material is covered, what is permitted to be done, and what additional conditions (if any) must be fulfilled. To determine these, preliminary investigations including a form of risk analysis must often be done and site visits to inspect premises where organisms or plant material would be held or handled are often necessary to ensure the existence and operation of adequate facilities and procedures. These licence conditions often need to be highly technical and closely matched to the cultivation or environment of particular species of plants or pest organisms. Therefore they should be formulated by appropriately qualified scientists.

Administration of import permits and licences

Different countries will vary in their willingness to issue import permits and licences, depending on their circumstances. In general, it is desirable to minimize the number of import licences issued. The first question for the potential importer, therefore, is whether the item could be sourced from a permitted area or from one less risky than that proposed. Very often it is found that, unbeknown to the potential importer, the desired item is available from a different supplier in a safe geographical area, or even in the home country, rendering the issue of an import licence unnecessary. It is also desirable to establish how necessary it is for the prohibited import to be obtained. This varies enormously and, in the risk analysis, is one of the factors that must be balanced against the risk of allowing the import. For example, there is a significant difference between a national need for a basic foodstuff and that of a private individual wishing to import an unusual variety of ornamental plant. The quantity of material desired is another important factor. Usually there are fewer risks associated with importing a small quantity than large amounts of any commodity. Finally, all import licences should specify in detail what must happen to the imported material after the purpose for which it was imported is finished. For propagation material of high health status, it may be acceptable to release it from control, possibly following a specified testing procedure, but for much material it is prudent to require destruction by some effective specified means. For convenience of administration, types of licences can be divided into those covering: (i) commercial material; and (ii) scientific or experimental material (including soil or growing media for plants, and genetically modified organisms (GMOs)).

Commercial material

While import permits continue to be required by many countries, import licences for commercial plant material sometimes may be needed because a commercial opportunity is perceived involving a prohibited item. There will be particular concern if the source area is a centre of origin or diversity for the plant genus concerned, as a whole range of pests attacking the genus will normally have evolved there contemporaneously, not all of which may have accompanied the plant in its spread to other regions. Nevertheless, by careful use of selected areas of production, inspections, tests and post-entry control, it may be possible to devise conditions that would reduce the risk of import to an acceptable level. It may also be possible to reduce the risk by verifying more carefully than normal the health of the material ancestral to that desired to be imported. Where the prohibition rests mainly on a lack of information concerning the plant and its pests in the country of origin, the supply of relevant information may clear the way for issue of an import licence, possibly leading subsequently to routine permission for import and inclusion in the national phytosanitary regulations. In any case, a careful risk analysis will need to be done, and the issue of an import licence should require a clearly positive conclusion with specified conditions.

Wild plants

When import licences are requested for plant collecting projects, it is usually not possible for the applicant to specify in advance exactly what material and which species it may be desired eventually to import. In these cases, consideration must be given to the nature and identity of the material sought and also whether there may be any conservation aspects. If the collection is to include vegetative material of important crop plants, the phytosanitary risk may be high. However, if the material collected is limited to true seeds, or the plant species are unlikely to harbour pests that might transfer to crops or to the wild native flora, the risk will be much less. The treatment and facilities for holding such imported material must be taken into consideration, and also any possibility of future commercial exploitation of the items collected. In most cases, it is possible to work out acceptable conditions. Dried plant specimens for scientific study normally present a negligible risk and seldom warrant control.

Scientific or experimental material

Licences to import prohibited harmful organisms or plants are often requested for scientific research, breeding and selection work, or for pesticide testing purposes. These are high-risk imports, especially because handling and some multiplication of the organism will probably be involved, and so there should be very good reasons given and safeguards provided before the issue of a licence could be contemplated. Even when the desired pest is not a major one, it is usually desirable to try to avoid the need for importation. Often it is possible and equally valid for the scientific work to use as its subject a related native species or a native strain of the same species. Alternatively, it may be possible to use an organism of similar biology but with a lower phytosanitary risk than that proposed. A similar approach can be used in relation to harmful organisms needed for pesticide testing. If the risk is considered too great for the work to be done in the country concerned, it is sometimes possible for it to be done in a country where the pest occurs naturally, or elsewhere where the risk would be much less.

Where issue of an import licence is contemplated, the facilities for containment of the pest or plant must be carefully considered and assessed (Chapter 12). For example, the risk of a harmful organism escaping from *in vitro* cultures in a laboratory is less than if it is held on plants contained in a glasshouse. The risk can be decreased, where appropriate, by creation of positive or negative atmospheric pressure within the growth room or glasshouse, in order, for example, to facilitate the exclusion of vectors from outside or the containment of the licensed organism within the structure, respectively. For invertebrate pests that are very difficult to contain, such as thrips, it may be necessary to isolate the chamber containing the pest within a cold environment, which would inactivate or slow down the activity of any that might escape. Containment facilities are considered in more detail in Chapter 12. A high standard of hygiene should always be maintained when dealing with licensed organisms, and a time limit and method for destruction at the end of their usefulness should be specified. Monitoring by the NPPO should be done during the period of licence validity to ensure adherence to licence conditions.

Soil samples

International movement of soil samples is often needed for physical, chemical or biological analysis. Soil samples for analysis are normally not difficult to handle safely, provided that any water used in their processing is controlled. Safe disposal of soil and other residues after analysis may present a greater problem if the analysis process does not itself cause sterilization. In this case, treatment with heat or a fumigant, and disposal by deep burial, is usually satisfactory.

Genetically modified organisms

Phytosanitary regulations often prohibit or restrict importation of GMOs, especially those that are plant pests, or where plant pests have been used in the modification process (the gall-forming bacterium, *Agrobacterium tumefaciens* is commonly used as a genetic vector). In the latter case, the degree of risk will usually depend upon whether the plant pest or its genetic components remain potentially infective or active. With others, it will depend on the nature of the modification. For example, a mechanically transmitted virus may be modified to render it transmissible by a vector, the host range of a fungus may be altered, or a plant may be modified to produce a toxin. The range of possible genetic modification is so vast and the techniques are so varied and are developing so rapidly that it is difficult to generalize. However, it is therefore obvious that applications for licences to import GMOs should be most carefully assessed by molecular biologists familiar with the techniques concerned, and that risky material should be tightly controlled if a licence is not refused.

Culture collections

Organizations that maintain collections of cultured microorganisms serve many purposes and sometimes have an extensive trade in imports and exports of cultures. They are frequently centres of taxonomic expertise and research, they provide reference material for diagnosis and they are sources of authenticated material for research, testing or industrial processes. Many countries have national collections of various types of microorganisms, including plant pests. Collections may be funded publicly, by a particular industry (such as brewing) or may be owned by commercial firms. Collections of cultured microorganisms are normally held *in vitro*, usually in a form suitable for long-term storage. This could be as freeze-dried material in sealed ampoules or as cultures under oil, deep frozen, or otherwise preserved in the living state. Some collections have a worldwide trade. Where national licensing laws operate, such culture collections will normally be appropriately licensed and monitored by the NPPO. New accessions need to pass through routine purification and identification procedures to ensure the cultures are pure and of the species and strain specified before adding them to the stock collection in a preserved state. Precautions against cross-contamination or inadvertent dissemination (Chapter 12) need to be particularly closely observed, but may not need any very special techniques. However, culture collection scientists and administrators in countries that are participants in the Australia Group need to be aware of export control restrictions, as described in Appendix III.

Invertebrates are not held in collections of living material as large as those for microorganisms. The difficulties of maintenance of living invertebrate cultures and the precautions needed to avoid escape or cross-contamination are usually greater, so collections tend to be smaller and limited to species currently being used for tests or research.

Import operations

Border entry points

Border entry points for the importation of goods may be on roads or railways, or at ports or airports. These points vary greatly in the volume of traffic passing through them requiring phytosanitary attention. Crossing points between neighbouring countries, especially on land borders, may be numerous and some may carry little traffic requiring attention. In such cases it may be reasonable to require international traffic subject to phytosanitary regulation to use fewer, designated, crossing points or to cross during designated hours. This permits closer control of such traffic, more efficient use of manpower and may save financial resources by the construction of fewer (but possibly better) facilities at BIPs. It may also be reasonable to limit the import or export of certain commodities to designated BIPs where facilities for handling them exist, but this must be technically justifiable. It must not be used as a means of limiting such trade, and it should not restrict competition between ports. Under the terms of the IPPC, it is necessary for participating countries to publish a list of such designated entry points, which must be selected in a way that does not unnecessarily impede international trade, and to notify the IPPC Secretariat, the relevant RPPOs and interested trading partners. BIP facilities are discussed in Chapter 5.

Staffing

Provided there is close liaison with the customs service, it may not be necessary for the NPPO to maintain a continuous official presence at all national entry points. Usually, the maintenance of continuous phytosanitary services can be justified only at entry points where the volume of trade subject to phytosanitary regulations is large. Where communications are adequate, it is normally sufficient for phytosanitary inspectors to be available on call at short notice.

Customs will normally be the initial contact for incoming transport. Where this is the case, an arrangement should be established whereby the NPPO is alerted by customs to any incoming goods that might be subject to phytosanitary regulations. Normally, these will be all goods accompanied by a phytosanitary certificate, which needs to be passed to the NPPO. However, there is also the possibility that the documentation for a consignment may be imperfect or the phytosanitary certificate may be missing, so customs need to be informed and made aware of all goods that might need phytosanitary control. Often the information for this can be incorporated in the customs computer system. Customs may also be in the best position to check the identity of the consignment against its accompanying documents, and have the authority to hold consignments for phytosanitary check and inspection by authorized inspectors.

Location of inspections

Inspections at the point of entry make it easy to take prompt and decisive action on faulty consignments (re-export, for example) and, provided consignments are uniform, large numbers can be examined. However, for the same reasons that make efficient inspection of goods in transit difficult, it is usually difficult and impeding to trade to attempt phytosanitary inspections at the point of entry. Without access to the whole consignment, it is very difficult to collect a representative sample for inspection. Therefore, unless phytosanitary inspections at the point of entry are essential because of the high phytosanitary risk of the material or the nature of the trade, it is usually more efficient to do them at the place of destination. Pre-clearance arrangements may also obviate the need for point of entry inspection, depending on the precise nature of the agreement between the trading partners. However, where inspection is deferred from the point of entry, it is necessary to make sure that consignments remain under the official control of the NPPO and that material potentially harbouring active pests is packaged in such a way as to make it unlikely that pests would escape en route to the place of inspection.

Import inspections at the place of destination have several advantages. Because the goods will all be off-loaded from the transport or emptied from their container, the phytosanitary inspector will be able to examine the whole of a consignment without having the major job of unloading and repacking the goods, and the resources this would have taken will be saved. Inspection can usually be done without major disruption of the work of the importer, as part of the unloading process. Also, the workload of import inspection can be spread amongst local inspectors within the country as well as those at the borders, with the potential of greater flexibility in the time when inspection can be done. However, in most cases it is prudent to make inspections as soon as possible after import, while the constituents of the consignment are still identifiable and, where appropriate, before the material is planted, marketed or otherwise disposed of. Occasionally it may be appropriate for an inspection to be delayed until the imported material has produced new leaves on which symptoms might be seen. An inspector responsible for a particular area must be alerted to the need for an inspection of imported goods by those at the point of entry, and will need to receive documents relating to consignments (or copies of them) before inspections are made. Local inspectors should know the regular importers in their areas and become familiar with their patterns of trade. A good relationship between importer and the local inspector makes it easier to do timely inspections and to keep each other mutually informed of developments in trade or phytosanitary regulations.

Action on faults (non-compliance)

Normally, it is not necessary or possible to inspect all incoming consignments that are subject to phytosanitary regulation. As explained in Chapter 7, decisions therefore have to be made on priorities and targeting of inspections, on sample size, and on sampling methods. Where faults are found within a consignment, the measures taken must be within the legal framework and appropriate to the phytosanitary risk. After detection, the first requirement is for accurate and authoritative identification of the pest or problem. Although inspectors will often be confident of their identification, it is usually advisable to send a sample for laboratory examination and determination, especially where there is any possibility of a legal challenge. This will necessitate holding the goods at a place and in conditions where there is little risk of pests escaping and where the goods will not deteriorate during the holding period. Temperature-controlled stores may be required for this purpose. When the identity of the fault has been confirmed, there will be a choice of possible measures to be taken. Action options are discussed in more detail in Chapter 7, but will range from no action (perhaps where the goods are unsaleable anyway, although they may still need controlled disposal) to immediate destruction of the most risky material. The range of measures will include:

1. Sorting and removal of affected items or plant parts.
2. Treatment of some kind, such as a chemical spray, or processing under certain conditions.
3. Permitted entry under official control of some kind, such as planting in quarantine for future testing and observation.
4. Re-export.
5. Destruction by specified means.

For legal purposes, and so that there will be no doubt about the action required or permitted, this advice and instruction must be given to the importer in writing as an official document or Notice. Difficulties sometimes arise when dealing with large quantities of material. For example, treatment of large amounts of timber by fumigation would require specialist attention with adequate equipment, which may have to be operated on site where movement of the consignment is impractical or too risky. Destruction of large quantities of material such as potatoes is also difficult, and it may be necessary to resort to deep burial in designated and officially approved waste disposal sites, with special precautions to decontaminate transport vehicles and prevent the possible escape of contaminated material en route (Chapter 12). In some circumstances destruction by industrial processing may be acceptable and will minimize financial losses.

As explained in Chapter 7 and in ISPM No. 13 (Anon., 2001c), notification to the exporting country and the relevant international organizations of the details of the non-compliance, together with mitigating action taken, is one of the more important associated tasks, which should be done promptly to help avoid further problems.

Post-entry quarantine

Holding imported material in quarantine for observation represents the only action where plant quarantine matches the traditional meaning of the word. Such material may be held in purpose-built facilities, such as official or private plant quarantine stations or, where facilities permit, in suitable conditions on the importer's premises. The facilities required and their operation are considered in Chapter 12, and detailed plans can be found in Kahn and Mathur (1999). Where imported material in quarantine is allowed to be held on private premises that are not already an approved plant quarantine facility, it is necessary to ensure that the required conditions are specified in the licence or permit, and that these are carefully checked before the material arrives and during the period of quarantine. Material held in post-entry quarantine should be inspected regularly to check for symptoms of quarantine or unknown pests, and may also be subjected to suitable tests to establish its health status. The period for which it is held must therefore be long enough to permit the appearance of symptoms, or for tests to be completed, before it is released or otherwise disposed of. Any charges for post-entry quarantine services will, of course, depend on the policy of the NPPO concerned.

Goods in Transit

Phytosanitary controls on goods that are not being imported into a country, but are merely in transit through it, should be minimal. Normally, it is sufficient to check that the necessary documents accompany the consignment and to verify that the goods appear to be those covered by the phytosanitary certificate and supporting documents. The consignment can then transit the country in customs bond. However, this assumes that the material is adequately packaged and sealed, and that it does not present a phytosanitary risk to the territory through which it is to pass. This is not necessarily so. If the material is not adequately sealed within covers and is of a kind that could harbour pests which could escape in transit, a more detailed inspection may be needed. For example, this could be the situation with raw timber still with bark attached, or with bulk consignments of potatoes. Risks can also be increased if the consignment has to be transferred between different forms of transport, or transferred from one ship or train to another. Circumstances often make a thorough inspection of consignments in transit difficult, except where change of transport necessitates unloading and the inspection facilities are suitable. Too often such inspections have to be done quickly, in cramped or unsuitable spaces or in poor light. There can also be external pressures on phytosanitary services not to interfere too much with goods in transit, as such trade can be profitable for the country through which it passes. Careful review of transit trade should identify the most risky items for targeting inspections, while risk analysis should provide guidance on how much effort and resources should be devoted to this work.

Exports

The basic procedure for issuing phytosanitary certificates for exports of plants, and other items for which they are required by the importing country, is set out in ISPMs Nos 7 and 12 (Anon., 1997b, 2001b). Legally, the NPPO should be the sole authority responsible for issue of phytosanitary certificates and should put in place a system for checking that exported goods conform to the phytosanitary requirements of the importing country, including any special additional requirements that may need to be declared on the certificate.

As explained at the beginning of this chapter, the wording of the IPPC model phytosanitary certificate places the onus on the NPPO of the exporting country to ensure that goods being traded conform to the phytosanitary requirements of the importing country. It is therefore essential for the exporting producer and trader to be aware of what these are. For the smaller and less-developed exporting countries, this can present a difficulty, as it requires considerable resources to keep up to date with all the individual phytosanitary regulations of a country's trading partners. One traditional way of doing this is to rely on the importing trader to provide information on the official phytosanitary requirements of his country for the goods concerned. This can work well and much trade is covered by this arrangement. However, very often difficulties arise because the importing trader has provided inaccurate or out-of-date information, or perhaps no information at all. This can result in heavy financial losses for the exporter, and sometimes for the importer as well.

Governments that are contracting parties to the IPPC or WTO-SPS are obliged to provide their phytosanitary regulations on request. However, this can take considerable time and the regulations may not be easily understandable. In recent years, it has also been possible to access the phytosanitary regulations of many countries through the Internet. The FAO and the local RPPO should also be able to provide information on the phytosanitary regulations of member countries in a language common to the exporting country. This is a great help, but in many cases there will still be a considerable amount of interpretation to be done in the exporting country, even when the language of the regulations can be easily understood. Such interpretation often rests on the local phytosanitary service, and can result in the expenditure of considerable resources in making enquiries and explaining to the exporting producer what tests or other procedures are needed to comply with the requirements. It is therefore helpful to all trading partners if phytosanitary regulations are written in as simple and as straightforward a way as possible, leaving as little as possible for local interpretation. This ideal is far from being achieved, partly due to the increasingly technical nature of the legislation, and some local interpretation will always be necessary in correlating local products, procedures and standards with those stipulated.

Information on foreign countries' phytosanitary import regulations

In countries with adequate resources, a part of the NPPO often supports exporters by preparing and issuing summaries of foreign countries' phytosanitary import regulations in the national language, and providing interpretations of requirements in local terms. The way in which this is done varies considerably, and the service can be free or chargeable, but there are some general principles that can increase the value of the summaries.

First, a standard format facilitates quick retrieval of the desired information. Through use, this format will become familiar to exporters, who will then know where to look for information on a particular product or requirement, such as what goods are prohibited from importation and whether an import permit or a phytosanitary certificate is needed for a particular item. Of course, this will require considerable initial work, particularly when the diverse formats of foreign regulations have to be edited into the standard format adopted. There may also be a need for language translation. But the result will be much easier for the exporter to use than a simple translation of the foreign regulations.

Secondly, it is helpful to provide interpretation of the foreign regulations in local terms. Regulations, especially after translation, are often not clear and the exact meaning may need investigation or interpretation. Sometimes clarification can be achieved by enquiry of the country concerned. Otherwise an informed judgement may need to be made, based on the known biology and attributes of the item concerned and the known situation of the importing country. For example, where planting material is required to be of a certain health standard, it is helpful to indicate (if appropriate) which grade or grades of stock certified by the exporting country would meet the import requirement. Again, if a requirement for virus freedom is made for a group of plants whose health status has nowhere been closely investigated, it could be taken that freedom from symptoms is what is meant. Experience with the requirements of each trading partner, as trade continues over the years, will gradually build up case histories and allow more authoritative interpretation in greater detail.

Examples of the general headings that could be used for these summaries are:

1. Origin of the information.
2. Definitions of unusual terms used.
3. Items, the import of which is prohibited, including lists of prohibited organisms.
4. Items that need an import permit.
5. Items that need a phytosanitary certificate and any necessary additional declarations.
6. Items that need a growing season inspection, and conditions for this, including pest-free period or zone radius.
7. Conditions relating to soil and growing media, either associated with plants or separately.
8. Conditions relating to plants for planting, including nursery stock, bulbs and corms, and material in tissue culture.
9. Conditions relating to cut flowers and foliage.
10. Conditions relating to fruit and vegetables.
11. Conditions relating to true seeds.
12. Conditions relating to wood and its by-products.
13. Conditions covering packing material.
14. Conditions covering consignments in transit or for re-export.
15. Permitted points of entry.
16. General information, including concessions for material in passengers' baggage or sent by post, and likely inspection or treatment action on arrival.

Export operations

Where a phytosanitary certificate is required for exported goods, a pre-export inspection will be necessary (ISPM No. 7; Anon., 1997b). This must be done as close as possible to the time of export to ensure that the health status of the goods does not undergo significant change before export, and the period of validity of the phytosanitary certificate should be stated. A period of 2 weeks before export is normally sufficient to allow inspections and any tests or treatments to be completed. Depending on the nature of the material, prior inspection of plants during growth in the field or nursery may be necessary (Fig. 6.1). For certain categories of material (for example, for hosts of fireblight, *Erwinia amylovora*) it may also be necessary to make inspections of the area surrounding that in which the plants are grown during the previous one or more cycles of vegetative growth to confirm the absence of a particular pest. For all these inspections, close liaison between the inspector, the producer and the exporter (if they are not the same) is essential. There may also be other requirements to check, such as the pedigree of the material or whether it has received some specified treatment. The producer and exporter, as well as the phytosanitary inspector, must be

Fig. 6.1. Plant health inspectors inspecting a narcissus bulb crop at flowering time. Such inspection may be for the purposes of bulb export, quality certification of bulbs as planting material, or both. The inspectors are checking flowers for trueness to the variety grown and the foliage for symptoms of disease or nematode attack. Note fork for lifting suspect plants. (Photo: author.)

aware of the authoritative phytosanitary requirements of the importing country in order that the required conditions can be met. Authoritative official summaries of foreign countries' phytosanitary import regulations, as described earlier, are therefore of great value for this purpose.

Records of phytosanitary certificates issued and related inspections, treatments, pest identifications, personnel involved and dates of activities should be kept for a reasonable period to permit trace-back in case problems are reported after certification. For this a period of 1–2 years is usually satisfactory. The system of record retrieval must be speedy and reliable, and electronic systems with security protection arrangements are most satisfactory for this purpose.

Re-exports

Where imported goods subject to phytosanitary regulations are destined for re-export, a phytosanitary certificate for re-export must be issued (ISPM Nos 7, 12; Anon., 1997b, 2001b). This will show the original country of origin. Where the consignment keeps its identity and is promptly re-exported, no inspection is needed, provided the original phytosanitary certificate satisfies the requirements of the importing country. However, where an imported consignment is repacked before re-export, possibly because the original consignment has been split up, amalgamated, or is held for a period longer than the validity of the original phytosanitary certificate, a pre-export inspection will be required. A difficulty sometimes arises where the country of destination has requirements (such as field inspection during the growing season) which cannot be met by the country of re-export. In these cases it may be possible to agree on suitable tests to be done or other alternative arrangements that would be acceptable to the country of destination. Otherwise no re-export certificate can be issued.

Special arrangements

There are various special arrangements which are sometimes operated to cover certain categories of exports.

Delegated inspections: grain exports

Where the volume of trade is large and the phytosanitary risk is low (for example, with exports of cereal grains to those countries requiring a phytosanitary certificate), arrangements are sometimes made for inspections and control to be done by trained inspectors employed by the industry concerned, under the authority and supervision of the NPPO. Such a system for grain exports starts with inspectors inspecting and sampling consignments before the grain is loaded into silos or on to international transport. At this point, representative grain samples can be examined on site and grain with live pests can be treated or rejected immediately. Otherwise, inspection, sampling and testing must be done during storage or at loading on to international transport. The grain transport containers and holds, and the silos or stores where grain is to be held pending export, must be free from infestation with relevant grain pests, so that pest-free grain is not subsequently contaminated. The NPPO will need to monitor these inspections and the condition of stores by means of spot checks and duplicate inspections. If live grain pests are found, the load should be rejected or placed in isolated storage pending fumigation or treatment with appropriate and officially approved chemicals. Such fumigation or treatment preferably should be done by trained professional operators. Rates of application and duration of treatments before the grain can be discharged will depend on many factors, particularly temperature, method of application and the species of pests present. Where required by the importing country, samples are tested prior to export at an appropriate laboratory or official seed testing station (OSTS) for the presence of regulated pests and weed seeds. The use of an automatic core sampler facilitates taking representative samples. The phytosanitary service is then notified of the results and, if appropriate, issues a phytosanitary certificate. Good records of inspections and treatments must be maintained. As with other feed and foodstuffs, additional tests for grain quality may also be necessary, but are not part of the phytosanitary control process.

Exports of seeds for planting may be dealt with in a similar way but, because volumes are normally relatively small, there is less reason for devolving export inspections to the seed trade. Here again, the OSTS will need to check samples for germination potential, weed seeds and other quality parameters, while a laboratory check is done, where required, for the presence of prohibited or regulated pests.

Pre-clearance of exports

In certain circumstances, for important trade commodities, the importing country may offer for its official representatives to inspect plants or other goods in the country of production prior to export, and to clear them, either provisionally or finally, for entry. This arrangement could infer a lack of trust between the trading partners and so is not universally encouraged. However, it can be beneficial to both trading partners and may be used, for example, to allow the exporting country time to install an acceptable inspection or testing system of its own. For the importing country, it provides confidence that the material has been produced and inspected to the standards it requires, while for the exporting country it provides assurance that exported material will not be – or is much less likely to be – refused entry, and thus will not risk serious losses by destruction or in transport and storage costs. Normally, such arrangements are only worthwhile for trade that is sufficiently great and mutually desired by both trading partners to justify the costs of stationing an inspector in the exporting country for the period necessary. It is usual for inspectors of the exporting and importing countries to work closely together, and there may be agreement for costs to be wholly or partly borne by the exporting country.

Communications and Statistics

The volume and speed of modern trade makes good communications and records essential. The use of a computer network for the administration of phytosanitary operations on imports and exports has become virtually essential, and needs to link inspectors, policy administrators and scientific support. Such a system has enormous advantages in providing good communication (by e-mail), on-line information from both within the country and internationally (for local instructions and foreign requirements), and as a means of recording the making and result of inspections, pest identifications and action, allowing the easy accumulation of essential statistics. It can also be used for countless other jobs, from holding lists and details of producers to automatically generating phytosanitary certificates and other documents from information provided. The planning, purchase and installation of such a system is a major exercise that must be done carefully, without too much hurry, and with close consultation between all parts of the NPPO and those responsible for the installation. Communications are considered more fully in Chapter 5.

Statistics

Accurate statistics provide the basis for good planning of phytosanitary services. Besides the volume and distribution of work, which will indicate the numbers of staff needed, statistics will provide information on the volume and fluctuation of various types of trade, faults detected, and the association of faults with particular sources, areas or suppliers. This, in turn, will help determine priorities for targeting inspections or monitoring, and provide information for PRA.

Eradication and Containment 7

Introduction

The maintenance of plant health by the active eradication or control of plant pests is a core activity of phytosanitary services, to which much of the other work done contributes. Phytosanitary action for the control or eradication of pests can only be applied after a pest has been detected. Most action can be divided into that required when pests are detected in association with imported material (interceptions) and that required when a new pest is detected as a replicating population (an outbreak). However, some types of action are applicable in both situations. Action can also be divided into short-term or longer-term measures, the latter usually being more appropriate for action on outbreaks.

Detection

Pests of phytosanitary significance come to the notice of NPPOs in many different ways and via a great variety of sources. Each case will need to be investigated and evaluated. For general gathering of information, it is helpful for NPPOs to foster good liaison with other areas of government, government agencies, and other organizations and individuals dealing with agriculture, plant pests and plant science, including universities, institutes, trade organizations, extension service personnel and consultants (ISPM No. 6; Anon., 1997a). Information on pest occurrence and spread may also come from international sources, such as RPPOs and FAO. Frequently, pests are detected as interceptions. This is the 'detection of a pest during inspection or testing of an imported consignment' (ISPM No. 5; Anon., 2002a). However, pests may also be detected during monitoring of intranational (domestic) trade. In this case, the material on which they are found may or may not itself have been imported, but is frequently related to imports. Sometimes pests come to notice when a grower seeks advice on an unusual symptom or requests diagnosis of a pest, and sometimes through notification by growers, importers or the general public. Specific surveys for pests will provide reliable information on pest occurrence. Traceback from detection or other information may lead to the discovery of an outbreak, which is 'an isolated pest population, recently detected and expected to survive for the immediate future' (ISPM No. 5; Anon., 2002a).

Notification to the national plant protection organization

Although many countries make it obligatory to notify the plant health authorities of the suspected occurrence of quarantine pests, this is seldom an important means of bringing them to official notice. In spite of publicity aimed at raising awareness of plant pests, growers and traders may not always be aware of a need to notify, and sometimes do not recognize quarantine pests or their symptoms. The NPPO becomes aware of serious pests more often via official inspections of one kind or another, through surveys, or via growers or traders seeking advice on a problem. For example, in the UK detection following notification is common only with the Colorado beetle (*Leptinotarsa decemlineata*), which is sufficiently conspicuous and distinctive to render it recognizable by most growers and the general public (although a large proportion of notifications from the general public turn out to be of harmless species). However, the obligation to notify is a helpful legal requirement when enforcement of phytosanitary measures is necessary.

Actions Related to Imports and Import Interceptions

Import inspections

Inspection of imported plants and plant produce should be methodical and should be done on a representative sample if the whole consignment cannot be inspected. As discussed in Chapter 5, for satisfactory inspection there needs to be adequate lighting, protection from the weather, including heating or shade where necessary, and facilities for handling the material. These may be no more than a table, but in the case of timber a crane may be needed, or a fork-lift truck where material is stacked on pallets. More specialized equipment may be needed for certain commodities. The contents of a consignment should be compared with the manifest, consignment note or invoice, and any visible discrepancies noted. During inspection, material should first be scanned overall, looking for potential faults. If present, inspections should include plants or plant products of stunted or uneven growth, unthrifty or unusual appearance, or other lack of uniformity. More detailed inspection should then be made. Some targeting of inspection on suspicious material may be appropriate. If no faults are suspected, then inspection is usually based on random samples up to a predetermined quota. Examples of visible pests, lesions, rots, damage and other symptoms should be examined closely, under the microscope if necessary and if possible. If quarantine pests, unidentified pests or prohibited material are found or suspected, then action must be taken and specimens preserved for confirmation of identification and future reference (ISPM No. 9; Anon., 1998b).

Targeting import inspections

Usually only a small proportion of consignments imported in trade can be inspected, even in countries with a large phytosanitary inspectorate, so it is advisable to target inspections carefully on those items that are considered to present a high risk. For example, such items may be commodities with a history of previous interceptions, or may come from areas with known phytosanitary problems, or which have been identified through PRA as posing a potential risk. A programme for targeting inspections, listing trade items or trade flows to be monitored, should be drawn up and revised regularly (at intervals of about 6–12 months) to reflect changing patterns of trade through the seasons and years. The order of priority for monitoring the various items should be indicated and, if appropriate, quotas should be set for monitoring inspections for each trade item for each period. Targeting will vary from country to country, according to the phytosanitary risk of commodities and trade pathways as assessed by PRA and according to the national importance of various crops. For example, plants for planting and propagating material would normally be considered a higher risk than produce for consumption. Where *Solanum* potatoes are an important crop, imported seed potatoes, or the crops grown from them, may be considered a high priority for inspection. Similarly, where tobacco is an important crop, items that might harbour tobacco pests, such as seed potatoes or tomato plants, would also be a high priority for inspection. In certain circumstances, for example, where large quantities of possibly contaminated waste may result, plant produce for consumption or processing may also present a high risk.

The best places for monitoring inspections are those where targeted material is normally being handled in the course of the marketing chain. These will therefore be national points of entry or destination for imports, nurseries and farms, processing factories, washing and grading units, packing and distribution points, and markets themselves. As mentioned in Chapter 6, it is normally unsatisfactory to attempt inspection of goods in containers unless these can be completely unpacked. Inspection is therefore often best left until arrival at the point of destination.

Notification of non-compliance and emergency action to be taken

When pests of phytosanitary significance are intercepted on internationally traded goods, or such goods are otherwise phytosanitarily unsatisfactory, this fact and the action taken should be notified immediately by the importing country to the exporting country (ISPM No. 13; Anon., 2001c). This notification should include all the necessary details, including the number and date of the phytosanitary certificate, the consignment reference marks and other reference numbers, to enable the NPPO of the exporting country to investigate the incident and apply corrective action. Contracting parties to the IPPC have an obligation to report the occurrence or spread of pests that are an 'immediate or potential danger' to other interested contracting parties (ISPM No. 17; Anon., 2002d). Any emergency action taken should be notified to interested contracting parties and to the IPPC Secretariat. Notification of the occurrence or spread of such pests and any action taken should normally also be made to the RPPOs of which the country is a member, and to the European Commission by EU Member States.

Where action is to be taken to control or eradicate pests, the owner of the land, crop or material in which the pest is found and the person in charge of the premises should be informed immediately and their cooperation sought. If these persons are not immediately contactable, a determined effort should be made to inform them, for example, by recorded post or by written notice given to a responsible employee.

Immediate action on interceptions

The primary aim of immediate phytosanitary action is to prevent further spread of the pest and to avoid the incident developing into a more serious outbreak or epidemic. The action taken will vary according to the circumstances, the nature of the pest or fault found or suspected, the nature of the material on which it is found, and on the phytosanitary regulations and policies of the country concerned. With many interceptions, the appropriate first action is to prohibit further movement of the material to prevent further dissemination of the pest. This also gives time for expert diagnosis of the pest and for an initial PRA to be made. Some kind of safeguarding treatment may be required, such as an application of pesticide. However, where there is no doubt about the identity of a serious quarantine pest, or there is imminent risk of further spread, immediate destruction of the affected material is often the most prudent course to take. This is frequently more economic than a protracted treatment programme, for which a successful outcome cannot be assured. The extent of the destruction necessary will vary and might not include the whole consignment, but could include other material nearby.

Contracting parties to the IPPC are entitled to take appropriate emergency action to counter the threat posed by a pest of phytosanitary significance, provided that the action is evaluated as soon as possible to ensure that its continued application is justified (ISPM No. 13; Anon., 2001c). National phytosanitary regulations should therefore enable emergency action to be taken, where necessary, on potentially serious unlisted pests or on suspicion, as well as on recognized regulated pests. Under the WTO-SPS such emergency measures must be appropriate to the circumstances and checked as soon as possible to ensure that they are technically justifiable, using PRA techniques. Within the European Union, precautionary measures to protect the Member States against such harmful organisms are obligatory under Articles 3 and 16 of Council Directive 2000/29/EC.

The detection of unlisted pests (Chapter 6), which have not been subjected to PRA previously, often causes problems. As outlined in ISPM No. 9 (Anon., 1998b), a PRA is essential both in deciding what action is necessary and to ensure that it is technically justified. However, any action should be taken as swiftly as possible to have the greatest chance of success, so there is usually very little time in which to do the PRA (Chapter 11). In many cases, therefore, the initial PRA may be in the form of an expert judgement. Where no precedents exist, judgement may be based on related examples and a wide knowledge of pests and control measures. More formal and documented PRA will depend on the available information, the extent of the measures to be taken and the level of justification needed. The more drastic the action, the more detailed the PRA generally needs to be. In some cases it may be necessary to do some research to supply essential information before a fully detailed PRA can be completed.

Inspectors must have confidence in their ability to recognize pests and their symptoms, and also in their ability to make judgements on appropriate action. If unfamiliar pests are found that appear to be potentially serious, action may have to be taken as a precaution even if identification is as yet uncertain. Although inspectors should not need to refer cases frequently, they should have ready access to more experienced or more expert opinion where necessary.

Short-term action options

The action options reviewed below are not comprehensive and cannot give many practical details, but they are intended to serve as a general guide to the kinds of action that are commonly appropriate to deal with interceptions. Any action must take account of the nature of the pest risk and the nature of the commodity in which it is found or suspected.

Prevention of movement

Where the fault with a consignment will take time to rectify or where there is suspicion of the presence of a quarantine organism on tradable material, normal practice is officially to prohibit the further movement of the material or to require it to be moved immediately into secure storage. Precautions against the escape of pests may also be necessary. This will allow time for further investigation and testing, where necessary. In some cases, where pests are highly contagious and can persist on the surfaces of machinery or containers (for example, with ring rot of potatoes, *Clavibacter michiganensis* ssp. *sepedonicus*), it may be necessary to restrict movement even within the owner's premises.

Additional documentation

Consignments of material subject to phytosanitary controls are frequently intercepted without the required phytosanitary certificates, with certificates that do not carry the requisite information or additional declarations on the certificate, or with goods that do not entirely conform to the description on the certificate. Within the European Union similar faults may occur in relation to the plant passports, which must accompany relevant material. In such cases, where no obvious phytosanitary risk is identified, the material may be held to allow contact with the NPPO of the exporting country and, where appropriate, the issue or amendment of the documentation by the issuing NPPO.

Re-export

This action is easiest to take when phytosanitary faults are discovered during inspections at points of entry. To be a practicable option, the integrity of the consignment must still be intact and it should not have been split up or partially sold, processed, planted or otherwise disposed of. It may also be appropriate when correction of faulty documentation is not forthcoming within a reasonable time, particularly if the consignment is of highly perishable material. This action should not be used if there is an immediate phytosanitary risk from the material in the consignment. Procedures are described in Chapter 6 and ISPM No. 7 (Anon., 1997b).

Sorting and removal

This is often appropriate where a quarantine pest is intercepted in trade, either at import inspection or during progress through the marketing chain. It is especially appropriate for produce such as fruit and vegetables when a relatively small proportion of the consignment is affected and the pest intercepted is very unlikely to spread to other produce or growing crops from the situation where it is found. For example, in the EU, sorting and destruction of affected fruits may be required when *Monilinia fructicola* is found on imported stone fruits. Most of the fruits allowed to be marketed will be consumed and thus destroyed. Also, this pest is unlikely to spread to orchards from points of entry, or urban shops or markets, especially outside of the local fruiting season.

Quarantine period

This means that the material in question is held temporarily in a secure environment for a specified period (see Chapter 12). This action is usually used for small quantities of imported planting or propagation material, or for exceptional and high-value plants. This may be a precondition for import, or to allow material imported during dormancy from an unreliable source to be grown on for observation and testing during growth. In other circumstances, it may be used as an alternative to destruction where valuable or rare material does not conform to requirements. The period of quarantine must be sufficient for any known quarantine pest likely to be associated with the commodity to cause visible symptoms or to allow reliable tests for pest detection to be done. Many countries have official quarantine stations with secure facilities (see Chapters 6 and 12) where such material may be held. However, quarantine may be carried out satisfactorily on private premises provided suitable secure facilities are available and their management is efficient.

Testing and monitoring of imports following an interception

Confirmatory tests are necessary in support of most phytosanitary action, especially when material is being held on suspicion of the presence of a serious pest or where there is any possibility that action may be contested in a court of law. Pest specimens should also be preserved and retained for a reasonable period (ISPM No. 6; Anon., 1997a). Laboratory tests and observation of material for detection and diagnosis of pests are an integral part of holding material in quarantine or under restrictions pending a decision. When suspicion of the presence of a pest is insufficient for immediate action to be taken, it may be appropriate to monitor the trade and check future consignments of similar type and origin. Detailed analysis of trade statistics may also reveal useful information. It is particularly important to follow up and monitor any planting material that has come under suspicion, especially if it has already been planted.

Treatment of interceptions

When faulty consignments with quarantine pests are intercepted, it may be appropriate to consider a treatment of some kind instead of re-export or destruction, or to support other action. This could be because the presence of a live quarantine pest is doubtful, it is one of lesser importance, infection or infestation is only slight, or the risk of the pest spreading is small and substantial loss might be avoided by applying a treatment. This alternative is more likely to be acceptable for produce or other items that are not planting or propagating material, although it could be considered in some circumstances, for example, where the planting material is particularly valuable or where producers are very largely dependent on imported material for propagation.

PHYSICAL PROCEDURES. These include trimming produce to remove organs that could carry the pest in question, and this may often be sufficient to reduce the risk to acceptable levels. For example, removal of leaf lamina from celery may satisfactorily ensure freedom from *Liriomyza* leaf-miners for marketing purposes. Pruning of pot plants or plants for planting may have a similar effect. In such cases, it is important to control carefully the removal and destruction of the discarded trimmings to ensure that pests do not escape in situations where they might find alternative susceptible host material.

Where the fault is soil on produce, it is sometimes practicable to remove it by brushing or washing, although washing without subsequent drying will usually cause the rapid deterioration of most kinds of produce. Unsterilized soil which may carry quarantine soil-borne pests, particularly nematodes, is a particular problem with certain types of goods, especially bonsai and penjing pot plants, which are traded as growing plants, sometimes in ornamental containers. These are often produced in small nurseries which may not always be aware of export requirements. Frequently, the subjects are the same species as, or are related to, those used in forestry or fruit production, and so may harbour the same spectrum of pests. They can also be very valuable items because good specimens take many years and much labour to produce. Indeed, the most valuable mature specimens may have been first planted long before modern phytosanitary requirements were formulated. It is therefore important to make sure that potential trading partners are aware of any restrictions on soil or on plant species commonly used for bonsai or penjing subjects before a trade in these items develops. Removal of contaminating soil and replanting plants in sterilized or soil-less media is very laborious and is seldom satisfactory, either for reducing phytosanitary risk or for saving the plants concerned, as mortality from this process can be high.

PESTICIDES. Chemical pesticide sprays, mists, fogs or dips can be used affecting the aerial parts of plants. If suitable pesticides which act systemically are available, these may also be used against pests attacking roots or other parts that contact sprays cannot reach. Systemic pesticides are also useful against pests such as scale insects and mealybugs which produce a coat of wax that protects them against contact-acting pesticides. It is essential that the pesticide used is officially approved for the purpose on the crop or species concerned. The pesticide and formulation must also be carefully selected to be effective against the stage of the pest present, while causing minimal damage to the host plant and the environment. The phytotoxicity of pesticides can be affected by many factors, including concentration and formulation, mode and duration of application, and age of plant, so an official requirement for pesticide application must exactly specify the various variables and should be a well-researched treatment that is known to produce reliable results. When there is no appropriate pesticide already approved for the purpose on the crop or species concerned, many countries have arrangements to extend the use of a product beyond those specified on the label ('off-label approval'). There may also be arrangements for temporary approval of materials for emergency application. Failure to follow pesticide regulations is a serious matter and may possibly result in the phytosanitary authority being liable for any losses consequently incurred.

FUMIGATION. In many cases, for example, where produce such as grain or root vegetables are being transported in containers, in railway wagons or by ship, fumigation of faulty material *in situ* may be possible. Fumigation is sometimes used as a routine precaution for the in-transit treatment of certain goods, irrespective of whether pests have been found, and can be very effective in controlling many different species and most stages of invertebrate pests. Good fumigants, applied in a proper manner, can penetrate quite densely packed materials, including soil, and give good control of pests that may be present within them. However, many chemicals used as fumigants are extremely toxic to mammals, including humans, and they must be applied with proper precautions and with the appropriate equipment. For this reason, it is usually advisable for all but the smallest official fumigation jobs to be done on contract by specialist firms or units. Large-scale fumigation of soil in the field or glasshouse must be undertaken with great caution and with close attention to possible risks to the environment, particularly the contamination of groundwater. Hitherto, the most effective and widely used fumigant chemical was methyl bromide, but this has been found to deplete the Earth's ozone layer and, as described in Chapter 12, it is currently in the process of being replaced by other materials, worldwide.

OTHER TREATMENTS. There are many other types of treatment that can be used to reduce the phytosanitary risk of faulty goods to acceptable levels. Dips in various pesticide formulations at ambient or higher temperatures, or in hot water alone, can be effective in controlling certain pests on produce or planting material such as bulbs (Dickens, 1979). For some invertebrate pests, a cold period is effective without damaging the host and this has been used satisfactorily as a routine pre-export treatment. For example, a 10-day period of cold storage at 1.7°C for glasshouse chrysanthemum cuttings imported to the UK from certain sources was effective in controlling *Spodoptera littoralis* infestations (Bartlett and Macdonald, 1993). Ware potatoes may be treated with a chemical sprouting suppressant, such as chlorpropham, to reduce the risk that they might be used for planting. Various kinds of irradiation, such as microwave, gamma ray and ultraviolet (UV) radiation have been tested for use as pest-control treatments for plants and other items subject to phytosanitary regulations (Sommer and Mitchell, 1986; Hallman, 1998). ISPM No. 19 (Anon., 2003c) gives guidance on this. Up to now the main phytosanitary use of irradiation has been for treatment of fresh fruit against fruit flies. Otherwise relatively few such treatments have been adopted, perhaps because of cost or the difficulties of treating large volumes of material. However, UV treatment of piped water supplies to control harmful microorganisms is an irradiation treatment that is quite commonly used.

Processing and controlled sales

Situations frequently arise where material comes under suspicion of harbouring quarantine pests but has not been specifically shown to do so by means of sampling and testing. This is usually due to its physical proximity to known infected material, but sometimes there are other reasons why the possibility of cross-contamination cannot be excluded. Sometimes treatment may be inappropriate or impossible, or it may be difficult to require the destruction of such material. This might be because of the large volumes of material or because phytosanitary laws do not allow compulsory destruction in the circumstances. In these situations phytosanitary risk may be reduced by allowing the material to be marketed under controlled conditions. For example, in the UK campaign against rhizomania disease of sugarbeet, beet from affected farms but harvested from fields where the disease had not been detected was allowed to go for processing to certain sugar factories that discharge their waste in a phytosanitarily safe manner, such as directly to tidal waters or, after water treatment, to watercourses not used for irrigation. Further precautions could also be taken, such as arranging for the beet from affected farms to be processed at the end of the season, after that from rhizomania-free farms has been finished, to avoid the interchange of mud on the wheels of vehicles serving affected and unaffected farms.

Other controls may limit the area in which the material may be used. For example, nursery trees suspected to carry *Plum pox virus* (sharka), but for which samples have tested negative, could be allowed to be sold for planting in amenity areas or in private gardens well away from areas of commercially grown hosts such as plum, peach or apricot orchards. Of course, in coming to such decisions, many factors need to be considered, particularly whether the aim is for eradication or containment of the pest, and its existing distribution and importance in the country.

Destruction

Where no other action will reduce the phytosanitary risks to an acceptable level, or where other action is impractical, then the affected produce or other material should be destroyed. To achieve maximum effect this should be done as quickly as possible and in such a way as to prevent the escape or spread of any pests the material may harbour.

Destruction by incineration is a preferred method because it eliminates the affected material and automatically kills any pests it may contain. However, the burning of large amounts of material is not always easy or safe, especially in the location where the fault has been discovered, and it is also environmentally undesirable. The material may have to be moved to another site where there are facilities for incineration, and in these circumstances consideration must be given to safety precautions and to the prevention of pest spread while the material is in transit. For much plant material or produce it is suitable to place it in plastic bags, which can then be sealed before movement and can be incinerated without prior opening. Where this is not possible, an effective pesticide treatment may need to be applied and allowed to take effect before movement. For very large volumes of material it may be necessary to move it in other sealable containers or in high-sided trucks, which are tightly sheeted over with impermeable covers. Every effort must be made to prevent pests escaping or contaminated material being dropped en route. The trucks and any containers, sheets, or other equipment that are re-used must be cleaned and disinfected after use.

It is not usually possible to incinerate large volumes of potatoes, root vegetables, or other very juicy or fleshy produce. In these cases, and in other situations where incineration is impractical, deep burial may be a suitable alternative. Burial should only be done in officially approved waste disposal sites (see Chapter 12), and material must normally be prepared for movement to such sites in the same way as described above for incineration. The buried material should be covered as soon as possible, preferably on the same day as the material is delivered to the site, and finally with the depth of soil necessary to prevent movement by birds, animals or the elements. This is frequently a legal requirement and is usually at least 2 m depth.

An alternative destructive process, which may sometimes be appropriate, is biodigestion by fermentation or composting in suitable bioreactors or composting facilities. Strict precautions must be taken to ensure the efficacy of the treatment against the pest concerned and, in addition, that there is no possibility of the pest escaping during the treatment period or through inadequate or non-uniform treatment of the contaminated material.

Action on Outbreaks

Short-term action

Pest outbreaks usually occur on crops or other host material in the field or in protected environments. Outdoor outbreaks may occur on hosts that are a constituent of natural vegetation, or which are planted in gardens or amenity areas, as well as on field crops. Appropriate immediate action will usually include measures corresponding to some of those already described for use in interception situations. In particular, the international notification of the occurrence of a pest outbreak of quarantine significance and emergency action to be taken is obligatory by contracting parties to the IPPC. The application of appropriate pesticides, fumigation of protected structures or soil, processing or controlled sales, and the destruction of crops or other host material may all be used to prevent further spread of the pest and to eradicate or contain the outbreak.

Pesticides and fumigation

The application of pesticides, even in emergency situations, should always conform to national regulations and to the manufacturer's instructions for their storage and use. However, in some emergency situations, where appropriate pesticides for the pest or crop have not been approved, it may be necessary to seek special authority for the temporary and limited use of appropriate pesticide materials. In many outbreak situations, the application of pesticides (especially fumigants) is better entrusted to specialist commercial contractors rather than staff of the NPPO, who may not have the appropriate specialist equipment or training. As described in Chapter 12, the use of the fumigant, methyl bromide, is being banned or severely curtailed by international agreement and any material contemplated for application as a space or soil fumigant must be carefully considered for its legality and effectiveness. In some outbreak situations soil treatment by solar heating (solarization) may be effective (Katan, 1981).

Crop destruction

Destruction of crops, especially root or tree crops, is often difficult. The extent of destruction should normally include areas assessed as very likely to be contaminated as well as the affected crop, part crop, host species, or area of vegetation. Sometimes it is appropriate to apply a pesticide treatment before crop destruction to prevent the escape of mobile pests, and particularly where destruction is delayed to permit a harvest to be taken. Destruction normally involves the application of an appropriate herbicide or desiccant at a rate that will give a rapid and effective kill of the crop or host plants. Trees will normally have to be felled and, if not uprooted, the stumps treated to prevent resprouting. If the pest is soil-borne it may be appropriate to apply a treatment to the soil after crop destruction. An officially designated controlled area surrounding outbreaks, of size depending on the mobility of the pest, is also usually destroyed, treated or otherwise regulated where appropriate, to act as a *cordon sanitaire*.

Longer-term action for eradication or containment

Contingency plans

Contingency plans for the elimination of serious pests that are anticipated to arrive at some time in the future should be drawn up before the arrival of such pests precipitates an emergency situation. This will allow a careful PRA to be done without pressure for a speedy decision on immediate action. This is especially necessary where there may be a need for international cooperative action, as this will inevitably require considerable time to negotiate, agree and put in place. Rehearsals of contingency plans should be done to identify and eliminate any problems during their development and to ensure that they will work well in practice.

Phytosanitary campaigns

If immediate action to eliminate the pest discovered is not successful, or a pest outbreak is already beyond the initial stage, longer-term campaigns for eradication or containment must then be considered. The detection of a seriously damaging pest outbreak affecting growing crops or wild native vegetation in a new area is a serious matter, and phytosanitary campaigns can be major and costly exercises. These may last months, years or, in the case of containment campaigns, may become a more or less permanent feature of the national plant health programme. Long-term campaigns will usually require the support of surveys and also, perhaps, special legislation and a designated budget. Before embarking on a phytosanitary campaign a PRA, including a cost:benefit analysis, should be done to provide a logical basis for the campaign and to ensure that it will be practicable and worthwhile (ISPM No. 9; Anon., 1998b). For large and extended campaigns a special management structure may need to be established, especially where such campaigns are international in extent.

Campaigns for elimination or control of a pest will vary according to many factors, particularly the type of pest, the host and the nature of the host environment. Action suitable for cultivated areas may not be possible or acceptable in forest or areas of natural vegetation. Also, the action possible and appropriate in protected or controlled environments may be impossible or inappropriate in field conditions. Many of the immediate action options discussed above may be used in the course of such campaigns, particularly pesticide and other kinds of treatments, and destruction of affected and suspect material. However, there are other action options that are more suited to longer-term action, and those most commonly used are reviewed briefly below. Normally, action will take the form of a package of complementary measures, usually including regular inspections or surveys for monitoring progress, destruction of affected and suspect plants and other material at some stage and frequently, especially for invertebrate pests, a programme of biological control. Publicity to foster public awareness, and consultation with stakeholders such as special interest groups and trade associations, will also be important to gain support for, and cooperation with, the campaign. Where import prohibition is to be avoided and repeated introductions of a pest render its eradication unsustainable, a policy of containment may be the only practical alternative, although it may still be prudent and practicable to attempt eradication where propagation material is involved.

Surveys and monitoring

Having secured the immediate situation by swift action to prevent spread from the outbreak discovered, it will usually be prudent to determine whether there are any other undiscovered outbreaks of the pest. For this it is necessary to do a survey, which is normally targeted on the particular hosts or habitat concerned. Outlines for survey and monitoring systems are given in ISPM No. 6 (Anon., 1997a). Surveys may also be limited or targeted in other ways to make detection more likely and to limit costs. For example, they may be limited to certain climatic, geographical or administrative areas, or to a certain time or stage of growth, such as flowering time, if flowers are the host organs affected or appear at a stage when symptoms are most apparent.

Monitoring of pest outbreaks is essential for assessing the progress and efficacy of the action taken (ISPM No. 9; Anon., 1998b). It is usually in the form of regular surveys or inspections, closely targeted on outbreak locations or on crops, areas, trade pathways, practices or producers that have some relation to the outbreak. It may also incorporate assessment and analysis of statistical data on the crop or trade affected.

Surveys may also be done for many other purposes, for example, to investigate the movement of pests in a certain trade pathway, to provide evidence of pest freedom for the establishment of a pest free area (ISPM No. 4; Anon., 1996d), or to justify the introduction of new phytosanitary measures. In cases where there is concern about the possible spread of a pest into new areas, surveys for a pest may be done before it is first detected and regular, routine surveys may be instituted annually or at more or less frequent intervals. Such regular surveys are needed where it is necessary to monitor closely the presence or absence or spread of a pest in a particular area or crop.

Surveys must be planned carefully in detail, including specific instructions to the personnel involved, how the fields or other areas to be surveyed are to be selected, arrangements and provision of equipment for taking samples and sending them for laboratory analysis or testing, and the policy and action on finding any further outbreaks in the course of the survey. The size, frequency and distribution of any samples to be taken must also be planned carefully in the light of statistical, practical and economic considerations. Before the survey is started, the organizers must be confident that the results will not be invalidated, or their value diminished, by poor statistical planning or lack of forethought on how the results will be analysed or on the practicalities of the exercise. This is especially important when the results may be used to justify action affecting international trade, to support the introduction of new phytosanitary regulations or to provide information for PRA.

Normally the results of a survey should be considered only when it has been fully completed. To obtain the maximum amount of information from a survey the results must be carefully analysed before conclusions are drawn. However, it may quickly become apparent that a pest is already firmly established and that further attempts at eradication are very unlikely to be successful. In these circumstances it may be decided to end the survey prematurely and to concentrate efforts on containment of the pest and minimizing the damage caused. The results of one survey often suggest reasons and modifications for the conduct of another, perhaps in the following season or at regular intervals.

Action suited to protected environments

Eradication of pests in glasshouses or other protected environments depends heavily on destruction of the affected material. Good hygiene and management are also very important in successful eradication campaigns. Infected or infested plants should be removed and destroyed, along with crop debris and other waste. Weeds should be kept to a minimum to avoid sustaining pests on alternative hosts, while an efficient programme of soil sterilization or use of commercial pest-free growing media should eliminate or avoid soil-inhabiting pest stages. A break in crop production permits thorough cleaning and treatment of both the structure and the airspace, using materials that may be phytotoxic or damaging to living plants, such as steam, chemical disinfectants and fumigants. The protected environment may also be manipulated to discourage the pest and assist in its elimination, with or without the presence of the crop. For example, the temperature and humidity may be held at the optimum for rapid completion of an invertebrate pest's life cycle, encouraging the emergence of adults, which can then be targeted with other treatments or traps (Cheek, 1999). The occurrence of the pest must be monitored, and eradication can only be claimed when the pest has been absent from traps for longer than the time taken to complete the life cycle. Light traps are particularly useful for monitoring lepidopterous pests and have been found to be more effective in protected environments than pheromone traps. Yellow sticky traps for most insects and blue sticky traps for thrips species are usually effective.

Eradication without crop destruction is particularly difficult where crop plants are present all the year round without a break, or where the presence of edible crops precludes the use of effective pesticides. Large, curtain-like sticky traps can be used to diminish pest populations in some circumstances, but not where they would also catch large numbers of parasitoids or predators in biological control programmes. For eradication or control campaigns in protected environments with all-the-year-round cropping, and where crop destruction is to be avoided, biological controls are an attractive option, sometimes as part of an integrated pest management programme using a package of measures applied in combination.

Biological control

Biological control of plant pests is a wide and complex topic, of which only a brief summary can be given here. Helpful accounts, among many in this context, are given by De Bach and Rosen (1991) and Hokkanen and Lynch (1995). The use of natural enemies to control serious pests has been used successfully on many occasions and in many countries. Classic examples include the control of the cottony-cushion scale (*Icerya purchasi*) of citrus in California by the predatory ladybird beetle, *Rodolia* (*Verdalia*) *cardinalis*, in 1888–1889, and control of the prickly pear cactus (*Opuntia* spp.) as an invasive weed in Australia by the larvae of the moth, *Cactoblastis cactorum*, during 1920–1925. There have been numerous examples since then, such as the control of cassava mealy bug in Africa by the parasitoid wasp, *Epidinocarsis lopezi*. These have mainly used invertebrate parasites or predators against invertebrate pests or weeds. Only in more recent years have there been cases where microorganisms have been used successfully as control agents, or where microorganisms have been the target of control (Navi and Bandyopadhyay, 2002). In most 'classic' cases the target pests were widely established and out of control before biological controls were applied. They had therefore lost the status of a quarantine pest and (in the more recent examples) had become the concern of the agricultural extension organizations rather than of the phytosanitary services.

The use of biological control agents in routine pest control has increased enormously since the mid 1980s, especially in protected environments. By the mid 1990s, pest control programmes for protected crops in the UK used biological control agents more extensively than chemical pesticides, and biological and chemical controls were frequently combined in integrated pest control systems (Cheek, 1999). The agents most frequently used at present are the parasitoid wasp, *Encarsia formosa,* and the predatory mite, *Phytoseiulus persimilis.* Many others are used elsewhere or in other situations. For example, the predatory beetle, *Rhizophagus grandis* has been effective in controlling the great spruce bark beetle, *Dendroctonus micans*, which is a quarantine pest in many European spruce forests (Evans and Fielding, 1996). Microbiological agents have also been used as biological pesticides, in which form they are usually subject to the same regulations as chemical pesticides. Various entomopathogenic fungi have been used in this way, usually as formulations of spores for spray application. For example, in protected environments with high humidity, *Verticillium lecanii* has been used successfully against aphids and other insect pests of ornamentals, while *Metarhizium flavoviride* has shown promise against locusts in southern Africa (Lomer and Prior, 1992). Entomopathogenic nematodes in the genera *Heterorhabditis*,

Plasmarhabditis and *Steinernema*, and the entomopathogenic bacterium, *Bacillus thuringiensis* var. *kurstaki* and the toxin it produces, have also formed the basis of other effective biological pesticides used against a wide spectrum of pests (Cannon, 1996). These biological pesticides and biological control agents have proved particularly useful for control in situations where there is serious concern about the environmental or food safety aspects of using chemical pesticides in phytosanitary campaigns, or where effective chemical pesticides are not known or are precluded from use by safety or other regulations. They are less useful against pests of crops with very short growing periods, such as lettuce, or in cool environments.

As routine biological control programmes normally aim to achieve a balance between the control agent and the target pest, such that the pest is maintained at populations below the threshold for economic damage, these are not usually appropriate for use where total eradication of the pest is the objective. Eradication of a pest by means of biological control cannot rely on the increase of the control agent by feeding or breeding on what will be a diminishing pest population. For eradication it will usually be necessary for inundative application of the control agent in numbers sufficient to overwhelm the pest population, often combined with other treatments in an integrated programme. For this purpose, biological pesticides and invertebrate biological control agents used in greater numbers can be effective. Entomopathogenic nematodes (*Steinernema feltiae* and *S. carpocapsae*) also have been used as inundative treatments in programmes for eradication of the sugarcane stem borer, *Opogona sacchari*, within the stems of woody ornamental house plants (Cheek, 1999). These and other nematodes are active against a wide range of pests, can be applied with conventional spray equipment, and have the power actively to seek out target pests as prey within substrates that chemical pesticides may not penetrate.

The arrival of new pests, with consequent efforts to eradicate or control them with various treatments, can seriously disrupt routine biological control programmes, and any new control programmes that are devised should attempt to minimize this disruption where necessary. The development of new systems of biological control usually require extensive research before they can be put into commercial practice. This can be costly and time consuming, so biological control is usually appropriate for use only where there is a major or widespread pest problem. Where there is no precedent, the identification and selection of a suitable organism for biological control of a pest may take a long time and requires very careful investigation and administrative control to ensure that it will not harm local crops, wild life or the environment. ISPM No. 3 (Anon., 1996c) sets out responsibilities and protocols for the import and release of exotic biological control agents. Even where a technique has been well developed elsewhere, it may need careful assessment and adjustment to local circumstances. However, successful biological control by a self-propagating organism can be remarkably cost-effective.

THE STERILE INSECT RELEASE TECHNIQUE. This biological control technique, sometimes employed for pest eradication or control in major plant health campaigns, was first used successfully for the eradication of the screwworm fly (*Cochliomyia hominivorax*), a pest of livestock and other warm-blooded animals in North America (De Bach and Rosen, 1991). Essentially, this technique involves the production and release of very large numbers of sterile male insects. Provided these are sufficiently numerous to form a substantial proportion of the total population of males and that the females do not normally mate more than once, a substantial proportion of the eggs produced will be infertile. Provided that the production and release of sterile males is maintained for a sufficient length of time, the population will gradually decline and will eventually become extinct. This technique can only be used when the situation and the biology of the pest meet many specific conditions and when the many problems of large-scale production of sterile male insects can be overcome. The technique is appropriate only where the pest causes regular and severe economic losses, where the pest exists in populations small enough to permit released males to form a substantial proportion of the population, and where there will be no immediate reinvasion from infestations elsewhere. Mass rearing must be cheap and there must be a reliable technique for selecting out males from the reared population. The pest must also suffer no other effects from the sterilizing treatment (normally irradiation), which would interfere with its normal mating behaviour, and the females should usually mate only once. Although few pest species satisfy these demanding criteria, for those that do the technique is very cost-effective, very specific and causes the minimum of ecological disruption. With plant pests it has been used successfully against the Mediterranean fruit fly (medfly), *Ceratitis capitata*, notably in Central America and the USA (Mitchell and Saul, 1990). It has also been successful against the melon fly, *Bactrocera cucurbitae*, on Kume Island, Japan, and the onion fly, *Delia antiqua*, in The Netherlands (Smith *et al.*, 1996).

The sterile male release technique is sometimes combined with the male annihilation technique, in which males are attracted to pheromone (sex attractant) traps and then destroyed. Male annihilation can be helpful in reducing the population of males to a level at which sterile male release can be effective, but it can also be effective when used alone.

Control of cropping

With soil-borne microbial plant pests that persist in the soil for long periods, it is very difficult and usually uneconomic to eradicate field infestations by means of chemical or physical treatments, although in warm climates soil solarization may be effective in decreasing infestation to levels that do not cause economic damage (Katan, 1981). There are many examples of such plant pests which have tough and thick-walled spores or other structures for survival of the organism, including wart disease of potatoes (*Synchytrium endobioticum*), red-core disease of strawberries (*Phytophthora fragariae*) and fusarium wilt of cotton (*Fusarium oxysporum* f.sp. *vasinfectum*). In these cases it may therefore be appropriate to control cropping on the affected land by officially designating its extent, location and the restrictions that apply to it. In the UK this is known as 'scheduling' of land. Although this does not actually eradicate the pest, except in the very long term (which, in the case of *Synchytrium endobioticum*, may be in excess of 40 years), this is a measure that can provide good control of the pest. To be effective, cropping controls must be operated under specific legal provision, the area must be accurately described and demarcated on a reliable large-scale map, and this written record must be safely retained and be locatable by the NPPO. There should also be some note made on the title deeds or land registry entry for the land in question so that new owners are made aware of any cropping controls or other restrictions that apply. Restrictions that are applied to such designated areas usually prohibit the growing of host crop species on the affected area, and it may be appropriate to require it to be grassed over to prevent movement of soil. They may also include less stringent restrictions in a designated buffer zone or *cordon sanitaire* surrounding the affected area. In these areas, for example, it may be appropriate to permit only resistant varieties of the host crops to be grown.

Use of resistant or tolerant crop varieties

Some ('tolerant') crop varieties are capable of becoming infected with, or harbouring, a pest to the same extent as normal varieties but without showing strong symptoms. Other varieties, which do not become infected with a pest as severely or as easily as normal varieties, are referred to as 'resistant'. Some ('immune') varieties may not be susceptible to infection at all. The decision on whether to permit the use of a crop variety on land infested with a soil-borne pest to which the crop variety is resistant or tolerant is often difficult. The cultivation of such varieties on the land area affected, at least while there is a high degree of infestation, may promote the development and selection of a resistance-breaking strain of the pest organism. Also, the use of a tolerant variety may encourage the persistence, and even the further spread, of the pest without increased infestation being apparent through the appearance of symptoms. The use of such varieties is therefore not helpful where such soil-borne microbial pests are the subject of official control or eradication campaigns. They are best reserved for use only after official controls have been lifted. On the other hand, the cultivation of immune varieties, which do not become infected by the pest, or resistant varieties, which do not permit the pest to multiply normally, may be helpful in controlling the spread of the pest, especially varieties that result in a decreased soil infestation by the pest after cropping. For example, in the UK the wilt-resistant hop variety Wye Target was found to reduce soil infestation by the hop wilt fungus *Verticillium albo-atrum* when grown on infested land (Chambers, 1985).

Cost:Benefit Analysis

In conducting an eradication or control campaign against a plant pest, it is logical that the cost of the measures taken should be less than the value of the benefits which ensue. An exception to this is where political considerations override scientific principles and for which no guide can be given. However, some kinds of benefit may be difficult to quantify in monetary terms, especially where the environment or some intangible benefit is concerned. Nevertheless, an attempt must be made to assess the balance of cost and benefit in equivalent terms in order to determine whether a campaign is or continues to be worthwhile (ISPM No. 9; Anon., 1998b). Many examples are given and discussed in Clifford and Lester (1988).

As with PRA (Chapter 11), cost:benefit analysis relies heavily on the availability and quality of data, particularly those for the relevant crop or trade concerned. Crop losses due to various pests may already have been assessed; several countries have programmes for regular assessment of losses to a wide range of pests. Where such measurements are not available, new research may be needed, but this requires time, which may not be available before decisions need to be taken. Extrapolation from measurements made elsewhere or on related crops or pests must be done with caution, because so many variables come into play that the information used may seriously distort results. However, in order to arrive at some conclusion, it may be necessary to make informed guesses to provide the input figures needed. Cost:benefit calculations are usually made on a national or regional basis, depending on the area of responsibility of the phytosanitary authority involved. It would obviously be inappropriate to make calculations on the basis of individual farms or holdings because those affected might well suffer large losses in the course of eradicating a pest for the good of the industry or area as a whole.

Costs of eradication or control campaigns vary enormously, but typically they include the pay and pension costs of the NPPO personnel involved, including a proportion of senior management, the cost of pesticides and equipment used, vehicle, fuel and travel costs, destruction and waste disposal costs, contractors' fees, cost of obtaining information, and (where this is done) cost of compensation payments for crops or produce destroyed. There may also be many kinds of indirect consequential costs. For example, in the UK during one campaign it was necessary to sink a new borehole to provide a replacement water supply to a school whose original water supply might have suffered contamination from soil fumigation treatments.

Simple direct benefits of pest freedom, such as increased yield of produce and savings on pesticides that do not need to be applied, are normally easy to quantify. Environmental benefits, especially where these are of an aesthetic or amenity nature, are often at least partially subjective and are much more difficult to quantify in monetary terms. There may also be a difference between short- and long-term benefits. Estimates of losses and benefits are often more durable and more widely appreciated if given in terms of proportions of the yield of the subject crop and not in monetary units.

Compensation

Many governments pay compensation to the owners of diseased or infested crops that are compulsorily destroyed in the course of a phytosanitary eradication or control campaign. This has several advantages. It reduces the hardship which such destruction may cause to the farmers and growers and tends to retain their support and cooperation for the campaign in hand. It also encourages them to report suspected new outbreaks by mitigating fear of the consequences. However, payment of such compensation can create serious difficulties for the NPPO and the national or local government.

In normal circumstances government budgets are never open-ended and are frequently tightly controlled, with very limited funds for phytosanitary campaigns. This will therefore inevitably cause difficulty when compensation payments reach the limit of the budget available. The choice then is either to exceed the budget, possibly taking much needed funds from other areas, or to cease payments and create discontent among those not receiving compensation. If the campaign develops into a major exercise, unlimited compensation payments can escalate to unforeseen proportions and create serious financial problems for government far beyond the concerns of the phytosanitary sector. There is also the problem of what to do about compensation for consequential losses. Destruction of obviously diseased or infested crops or material seldom raises objections, but it may also be essential to destroy symptomless plants that may possibly be harbouring the pest. The extent of such destruction is often a partially subjective judgement and may involve very much more material than that which is shown to be infected or infested. In addition, farmers and growers may suffer losses due to the costs of destruction or other treatment, or because of lost sales due to restrictions imposed. Governments are usually much more reluctant to compensate for these losses as this would often rapidly exhaust the budget or render the campaign uneconomic.

The amount of any compensation payment is also sometimes difficult to determine. The usual aim is to compensate at the market value. However, markets often fluctuate considerably and if the crop is immature its volume and quality at harvest will have to be estimated, taking into account the inputs that will not have to be made. Too complicated a calculation for determining compensation should be avoided as it frequently leads to errors and high administrative costs. The result will not satisfy all growers, who will remain dissatisfied if the figure appears too low and who might be tempted to encourage spread of the pest if it is too generous.

An alternative policy is for the government to pay no compensation at all for losses incurred by farmers and growers as a consequence of phytosanitary eradication or control campaigns, but to regard any losses as an integral risk of crop production. In this case it would be appropriate for farmers and growers to take out private insurance against such losses, or for the relevant industry to act as its own insurer by maintaining a levy-funded compensation fund. This arrangement would increase production costs, but if done on a sufficiently large scale the increase would be small and the costs would fall appropriately on the industry's clients and not on taxpayers in general.

International Campaigns Against Migratory Plant Pests

Certain notable plant pests are migratory, breeding in one place and moving, sometimes long distances, to attack vegetation in another. Such pests include locusts and grasshoppers (which were recorded as plant pests in biblical times), various armyworms (notably the larvae of certain noctuid moths), and several species of grain-eating birds, particularly those in the genus *Quelea*. Campaigns against these plant pests are not usually the responsibility of the NPPO, but are normally dealt with by a different part of government administration for agriculture. However, occasionally the same official personnel may be involved with these as with other plant protection and phytosanitary campaigns. A brief summary of this topic is therefore included here.

Locusts and grasshoppers

These pests all belong to the family Acrididae (short-horned grasshoppers) in the Order Orthoptera (Steedman, 1990). The principal pest species are shown in Table 7.1. Of these, the first two are globally the most important. Although each species differs in its biology and ecology, true locusts are characterized by existing in two phases. In the solitary phase they inhabit relatively small 'recession' or 'outbreak' areas in which a sparse population is maintained. When the climate is particularly favourable for breeding (ample rainfall, allowing easy egg-laying and resulting in abundant vegetation) their numbers increase and, at a certain point, this triggers transformation into the gregarious phase, in which there is a tendency to aggregate into dense bands of flightless immature hoppers and swarms of adults. Further breeding over several seasons may result in immense swarms covering many tens of square kilometres. These swarms migrate, travelling down-wind up to several thousands of kilometres. In the case of the desert locust, this results in swarms accumulating in the inter-tropical wind convergence zone

Table 7.1. Principal pest species of locusts and grasshoppers.[a]

Species	Distribution
Migratory locust, *Locusta migratoria*, with several subspecies	Western, central, eastern, southern and south-eastern Asia, southern Europe, northern and sub-Saharan Africa, Madagascar, Australia
Desert locust, *Schistocera gregaria*	West Africa eastwards to south-west Asia
Red locust, *Nomadacris septemfasciata*	Central and southern Africa
Brown locust, *Locustana pardalina*	Central and southern Africa
Italian locust, *Calliptamus italicus*	Southern Europe and North Africa to central Asia
Moroccan locust, *Dociostaurus maroccanus*	Mediterranean, and western Asia
Australian plague locust, *Chortoicetes terminifera*	Australia
Argentinian locust, *Schistocera cancellata*	Southern South America
Senegalese grasshopper, *Oedaleus senegalensis*	West Africa to south-west Asia
Variegated grasshopper *Zonocerus variegatus*	West, Central and East Africa

[a] About ten other species of grasshoppers are of occasional importance in West Africa and some other *Schistocera* locusts sometimes cause problems in South America.

(ITCZ) from Mauritania to India between June and October. Those moving southwards stay within the ITCZ, while those moving northwards do so on southerly winds associated with depressions moving eastwards through the Mediterranean and western Asia. Grasshoppers do not exhibit the solitary and gregarious phases of locusts and, although they may become very numerous, they normally remain only a local problem.

Strategies for control were developed during locust outbreaks from the 1930s to 1960s (Krall *et al.*, 1997). Good control of locusts depends on early detection and elimination of hoppers before flying swarms are formed. At first, arsenic baits were used, followed in the late 1940s and 1950s by the persistent organochlorine insecticides, BHC and then dieldrin, which remained active on sprayed vegetation for many weeks. These were applied in swaths to kill bands of hoppers before the adult swarming stage was reached. Hoppers crossing these swaths accumulated lethal doses of insecticide by contact or ingestion. By the 1950s and 1960s control depended heavily on dieldrin applied in ultra low volume (ULV) formulations with rotary atomizers. However, in the more developed countries it became apparent that the organochlorine pesticides were having severely deleterious effects on the environment and other organisms, and this led to a ban on their use in most countries, which also resulted in preventing their use for locust control.

In later outbreaks, the less persistent organophosphorus, carbamate, and synthetic pyrethrin insecticides were used, both for hopper control and for emergency treatment of flying swarms by aerial spraying. Whereas dieldrin had been applied only once in each swath, the less persistent contact insecticides such as fenitrothion necessitated not only much more extensive applications but, where successive bands of hoppers appeared in the same areas, applications frequently had to be repeated. This also resulted in harmful side-effects on non-target organisms and the environment. Other control methods were therefore sought. In the 1990s good results were obtained with biological control using myco-pesticides based on the entomogenous fungi *Metarhizium flaviviride* and *Beauveria bassiana*. There have also been promising results with insect growth regulators that interfere with the development and metamorphosis of immature hoppers to adult locusts, and with the phenyl pyrazole pesticide, fipronil. This is relatively persistent on vegetation and in soil but does not bioaccumulate like organochlorines. However, more studies on food-chain effects are needed.

Early detection and elimination of hoppers is not an easy task in many of the countries affected. Regular meteorological observations and entomological surveys are needed in the recession areas to detect favourable weather conditions, egg laying and the emergence of hopper bands. However, finance is often very insecure and, in Africa, depends largely on donor agencies. There is a tendency to run down both funds and experienced locust control teams during recession periods when locusts are not a problem. Also, the recession areas in which hopper bands first form are often difficult to access because of their remoteness, poor transport, or military conflict. Nevertheless, no famines have been attributed to locust plagues during the past 50 years, although individual growers have suffered severe losses, and in recent years there have been serious doubts about the cost:benefit value of desert locust control in Africa, partly because operational difficulties there make locust control unnecessarily expensive. By contrast, in Australia, where most of these problems can be solved, it has been shown that good control can be regularly achieved with efficient organization and a moderate but secure budget.

Armyworms

These are the larvae of several species of noctuid moths, which are so called because of their habit of congregating together in vast numbers. Damage by these pests to graminaceous crops, including rangeland pastures, cereal crops and sugarcane, occurs in tropical and temperate climates on all continents. The adult moths, which are weak nocturnal flyers, are migratory to varying extents and are carried overnight on the prevailing winds, sometimes for several nights consecutively, up to several hundred kilometres distance.

Mythimna unipuncta is the main species of armyworm in North and South America, Europe and, to a lesser extent, in Africa, while *M. separata* (oriental armyworm, with which it was formerly confused) is the main species in Asia and Australia. In Africa the main species are *Spodoptera exempta* (African armyworm) and *S. exigua* (lesser armyworm). An excellent account of biology and control is given by Meinzingen (1993). As with locusts, *S. exempta* occurs in solitary or gregarious forms, the larvae of which differ in colour from the fourth instar onwards. There are six larval instars, most damage being caused by gregarious larvae of the fourth to sixth instars. Favourable climatic conditions (onset of rains after a dry season) facilitate breeding and survival, and the resulting high population of larvae triggers gregarious behaviour. Adult moths of later generations migrate and further outbreaks occur where egg-laying moths have been concentrated by the wind currents.

As with locusts, accurate monitoring of armyworm populations permits good forecasting and control. Adult moths are monitored by use of light or pheromone traps, the latter having the advantages of not requiring electricity or a skilled operator to identify the catches because they generally catch only mature male armyworm moths. Trapping is supported by surveys for eggs and larvae when conditions are favourable, but samples of these must be sent for identification to the national coordinator (see below), to whom trap catch data must also be sent. This information, together with knowledge of the regular seasonal migrations and prevailing environmental conditions, is used to formulate and issue regular national and regional forecasts of armyworm infestations. Control of larval stages is with synthetic pyrethroid, organophosphate or carbamate insecticides, using standard or ULV spray equipment, which may be mounted on vehicles or aircraft to treat very large infestations.

Bird pests

Many bird species can cause local damage to small-grain crops (grains other than maize), but various species of *Quelea* can cause serious damage to these crops in most countries of sub-saharan Africa. Although they do not migrate over such long distances as locusts, migrations have been known to cover 1000 kilometres or more, and frequently cross international boundaries. The queleas are small, sparrow-like birds in the family Ploceidae (weaver-birds). There are three species with overlapping ranges (Irwin, 1989), as shown in Table 7.2.

Of these the most important is the red-billed quelea, *Q. quelea*, of which there are several sub-species. The birds breed in huge colonies in dry thorn-bush areas and in non-breeding periods roost at night in dense flocks. Vast flocks of several million individuals can form. Food consists of various grass seeds, supplemented by insects during breeding. When grass seed becomes scarce, after it has germinated at the beginning of the rains or through being hidden under vegetation later on, the birds migrate to areas where grass seed or small-grain crops are available, often breeding immediately on arrival. Flocks often migrate again after breeding and birds bred in one country may cause damage in another (Ward, 1971).

Control has been by destruction of breeding colonies, using explosives to ignite mixtures of flammable petroleum liquids and by attacking the birds at roosting sites with poison sprays (Mundy and Jarvis, 1989). Breeding colonies are often inaccessible for the same

Table 7.2. Species of *Quelea* and their distribution.

Species	Distribution
Q. quelea (with several subspecies)	Sahel, eastern, central and southern Africa
Q. erythrops	Wetter areas of west, central and south-east Africa
Q. cardinalis	Eastern central Africa

reasons as locust recession areas. It is not possible to predict where flocks will go after breeding, and destruction of colonies seldom decreases crop damage substantially. Control efforts are therefore more usefully concentrated on roosts that immediately threaten areas of small-grain crops. 'Trap roosts' of planted vegetation such as sugarcane or napier grass (*Pennisetum purpureum*) are often more attractive to queleas than natural vegetation, and can be sited in accessible areas where collateral damage can be minimized. Another advantage is that they tend to harbour fewer birds of non-target species and so decrease mortality of these. Chemicals used for quelea control have included the organophosphate pesticide parathion, which is cheap and widely available but extremely toxic to all vertebrates as well as insects. Another organophosphate, fenthion, is equally toxic to small birds, but very much less so to mammals, and adjustment of the application rate can also avoid killing larger birds. Although more expensive, this is therefore preferable to parathion. Application to roosts has been by both aerial and ground-based spraying of the chemicals formulated in diesel fuel. Spraying at dusk or in darkness may be necessary to avoid disturbing the birds and to ensure that the maximum number are concentrated in the roost. Aerial spraying at these times can be hazardous and ground-based operations using a modified mist-blower have been found cheaper, easier to organize, and give good results.

International cooperation

Locusts and birds may do little damage in countries in which they breed, so it is tempting for these to overlook responsibility for swarms or flocks issuing from their territory. Nevertheless, these pests frequently migrate across international boundaries, so international cooperation is necessary for their control.

International regional organizations have evolved mainly for locust control, although control of armyworms and bird pests is also sometimes included in regional activities. From about 1930 onwards locust control activities, including monitoring and forecasting, were coordinated by the Anti-locust Research Centre in the UK. The Anti-locust Research Centre also maintained an archive of locust control literature and reports from 1929 onwards, and this is retained by its successor institutions (currently the University of Greenwich). This work was taken over in 1978 by the FAO, Rome. Organizationally, migrant pests are dealt with by the Locusts and Other Migrant Pests Group within the FAO Plant Production and Protection Division's (AGP) Plant Protection Service (AGPP). This is supported by Plant Protection Officers stationed at FAO regional offices and the Desert Locust Control Commissions. The Migrant Pests Group coordinates international regional efforts for eradication or control of locusts, African armyworms and *Quelea* and, where necessary, also coordinates contact between donor organizations and affected countries (website: www.fao.org/news/global/locusts/locuhome.htm). It issues early warnings based on reports received, and assists action, including contingency planning and emergency preparations. In forecasting locust development, the FAO Desert Locust Information Service uses the Schistocera WARning Management System (SWARMS), a geographical information system (see Chapter 11) developed for the purpose.

Several international regional organizations are concerned with coordination and facilitation of locust control. The Desert Locust Control Organization for Eastern Africa (DLCO-EA) has seven member states and is located in Addis Ababa, Ethiopia. In West Africa the regional organization covering nine countries of the sub-Saharan sahel is the Comité Permanent Inter-Etats de Lutte contre la Sécheresse dans le Sahel (CILSS). Its training arm, the Départment de Formation en Protection des Végétaux, established in 1981, gives training in crop protection, disseminates information and promotes integrated control methods. Other regional organizations in West Africa were formed separately to combat bird pests and to control locusts and grasshoppers. These were respectively the Organisation Commune de Lutte Anti-Aviaire and the Organisation Commune de Lutte Anti-Acridienne. In 1965 they combined to form OCLALAV, with ten member states. Further south, the International Red Locust Control Organization for Central and South Africa covers the southern half of the continent. For successful operation, each of these organizations, including the FAO, depends heavily on the cooperation and action of its member governments and on their national locust and bird control units to carry out surveys of their territories and to feed back data on which forecasts and action recommendations can be based.

Principles of Certification and Marketing Schemes 8

Introduction

Certification and marketing schemes in the agricultural context are administrative systems for quality control of propagation and planting material. They can be applied to production both of true seed and of vegetative propagation and planting material for crops normally propagated in this way. They provide quality assurance for the purchaser, while the added value imparted by independent quality assessment should increase rewards for the seller. Essentially, such schemes provide for official inspection of crops producing propagation or planting material, provide for checks on health, vigour and conformity to the characters of the relevant variety, and award certificates or labels to material that successfully meets the set standards and scheme regulations. This is normally referred to as 'certified' material. Schemes may be voluntary or compulsory, and within a scheme there are often several different grades of material with more or less stringent standards according to grade.

There is a considerable difference between schemes designed to cater for vegetative propagation material, such as seed potatoes, bulbs, or the mother trees, scions and rootstocks, which are used to produce young fruit trees, and those for true seeds, such as for cereals or oilseed rape. Although there are a considerable number of serious seed-borne pathogens, these are relatively few compared with those transmitted in other ways. Partly because of this, seed schemes tend to be more concerned with genetic purity and other aspects of seed quality, such as germination potential and freedom from weed seeds, than with health. True seeds are normally the product of sexual reproduction and they therefore have a greater potential for genetic variation than does material propagated vegetatively, and the seed schemes therefore also tend to be relatively more concerned with conformation to variety characters than schemes for vegetatively propagated material. However, vegetatively propagated crops are prone to acquire and accumulate pathogens from generation to generation as chronic infections that are often systemic, and their health status is therefore relatively more important. Because of these differences, schemes for true seeds and those for vegetatively propagated crops are often dealt with by different parts of the government administration responsible for agriculture, and in this chapter they are described separately.

There is a large but scattered literature on crop certification. Rudd-Jones and Langton (1986) cover many aspects, while certification for seed potatoes is dealt with by Shepard and Claflin (1975), certification of pome fruits in many countries by Rosenberg and Aichele (1989) and Ebbels (1989), and certification of grapevine by Martelli (1992). EPPO has devoted considerable resources to the preparation and publication of definitive guidelines or 'blueprints' for schemes covering many fruit and ornamental species grown in Europe (see Chapter 9).

Terminology

Schemes for vegetatively propagated material that has a pedigree or 'filiation' requirement that the certified material is derived from pathogen-tested original plants (the 'nuclear stock', see below) differ fundamentally from those that do not have this requirement and pass or reject material mainly on a visual inspection during the growing season. In the EPPO guidelines mentioned above, the term 'certification scheme' is used more specifically to denote schemes with a pedigree requirement, those without this requirement being referred to as 'classification schemes'. Elsewhere, schemes without a pedigree requirement are sometimes referred to as 'marketing schemes'. But 'marketing schemes' can also refer to systems using marketing regulations, which do not cover conditions for growing crops but specify standards and procedures simply for marketing plant material. Another source of confusion is that in Member States of the European Union, the certification process for seed potatoes is often referred to as 'classification', to denote the placing of stocks into the grade classes of 'basic' (intended for the further propagation of more seed potatoes) and 'certified' (intended for the production of ware potatoes). There is thus a risk of confusion in the use of the term 'classification scheme'.

These terms are not definitive and all types of schemes that issue certificates or labels to material that satisfies the scheme conditions can be referred to as 'certification schemes'. However, these are clearly distinct from, and must not be confused with, export certification resulting in the issue of an international phytosanitary certificate (Chapter 6).

Development

As is often the case in the history of plant pathology, the potato played a key role in the development of quality control schemes, the early history of which is outlined in Chapter 2. As described, a system of inspection, evaluation and certification for seed potato crops was established in Germany in the early years of the 20th century (Appel, 1915). The concept of certification for seed potatoes was quickly appreciated and promoted by W.A. Orton and others in North America. At the fifth annual meeting of the American Phytopathological Society in Atlanta, Georgia, over the New Year of 1913–1914, Orton proposed a system of official inspection, organized by state government agricultural authorities and supported by state legislation, for which entry of crops would be voluntary and for which the grower would pay (Orton, 1914). He was also careful to propose the criteria that would have to be met by those crops receiving a certificate. These covered 'uniformity to type' (conformation to variety), freedom from certain 'dangerous' diseases (listed as 'powdery scab, wart disease, eel worm, fusarium wilt, verticillium wilt, southern brown rot and leafroll'), and a limit on mixtures with other varieties. There were also limits on the permitted incidence of other diseases such as '*Oospora* scab' (skin spot), *Rhizoctonia*, silver scurf and 'curly dwarf'. These criteria, with some modifications to keep pace with the advancement of science, still cover the basic principles for seed potato certification today.

The idea of official certification for quality control of seed potatoes quickly spread throughout the world. In Britain certification schemes were first introduced for seed potatoes in 1918, but the emphasis at first was on conformation to variety and not on freedom from disease. The reason for this was the need to control potato wart disease (*Synchytrium endobioticum*) by the use of immune varieties. Previously, there had been much confusion over the names for potato varieties and, once this had been clarified, seed potato certification provided authentication, so that growers purchasing seed potato stocks could be confident that they were true to variety (Ebbels, 1979).

The success of schemes for seed potatoes encouraged their application to other crops that had similar needs for disease control and variety authentication, and they were applied both to other vegetatively propagated crops and to crops propagated from true seed. From the 1930s onwards, rapid progress was made in plant pathology, particularly plant virology, and the pathogens causing many types of symptoms were determined. This increased the emphasis on health status in schemes for vegetatively propagated crops, until health became of equal or greater importance than authentication and purity.

Many species of crop plants give a greater yield of improved-quality produce when they are free from detectable harmful organisms, sometimes even when no overt symptoms are apparent. Stocks of vegetatively propagated crops, in particular, readily accumulate infections with systemic pathogens, because such organisms will be present in most parts of the plant including those forming the propagation material (the seed potato tubers or young strawberry plants produced from stolons, for example) and are thus transmitted from generation to generation. The gradual decline in health and vigour of such crops observed in earlier times is now known to be due to this reason and to the subsequent multiplication of the infected individuals to produce greater numbers of infected progeny. It is also possible for other harmful organisms to be disseminated on propagating material without causing systemic infections if they habitually attack the propagative plant organs, or because they inhabit the soil adhering to the planting material. It follows that if a nucleus of healthy plants can be obtained, and provided re-infection by the relevant pathogens can be substantially prevented by means of a certification scheme, such stocks can be multiplied and yet kept healthy for an indefinite period. Where re-infection cannot be controlled, or the harmful organisms do not spread mainly via the propagating material, the harmful organisms are not amenable to control by certification systems. Generally, aerially dispersed organisms and those where early infections are not easily recognizable lie in this category.

Nuclear Stock

Schemes that have a pedigree requirement (that material certified must have been propagated from material itself previously authenticated or certified) start from an initial plant or plants that have been authenticated by examination and testing as true to variety and of the health status desired. Such plants are termed the *nuclear stock*. In schemes designed to produce very high quality material, the normal aim is for the nuclear stock to be free of all known pests affecting the species and crop in question, particularly pathogens that could be transmitted in or on the planting material.

Methods for producing pathogen-tested nuclear stock were developed from the 1930s onwards and are still being refined as new tests and methods of pathogen detection come into use. For any species or variety of plant, the basic requirements are to detect and identify pathogens present, which could be transmitted with the propagating and planting material, and then to select or produce material free from detectable pathogens. In many cases, these pathogens are viruses or organisms that behave in a similar way. The simplest method is to select plants that are already free of the relevant pathogens. However, if the available crop plants are all infected, this will not be possible and material must be freed from pathogens using special techniques.

Early tests for virus infection relied on symptoms displayed by inoculated indicator plants. These tests had certain advantages but were slow and cumbersome to perform. These are now complemented by modern tests using serological or molecular methods, which are rapid and specific (Chapter 10). Heat therapy as a means of freeing plants from virus infections was developed in the early 1950s for strawberries (Posnette, 1953), and was later extended to tree fruits and other species (Hollings, 1965) in conjunction with tissue culture (see Rapid methods of plant propagation, p. 158) to propagate from the plant organs thus rendered virus free. Other methods, including the use of cold therapy for freeing plants from infection by viroids and chemotherapy against viruses (Fridlund, 1989) and phytoplasmas (Jones, 2002), have also been effective. Many specialized methods for freeing plants from infection by numerous other different kinds of harmful organisms, including fungi and fungus-like organisms (e.g. Hirst *et al.*, 1970), bacteria and nematodes (e.g. McNamara and Cleia, 1985), have been developed.

Fig. 8.1. Propagation of high grade certified strawberry plants in a field well isolated from other strawberries. Note the wide spacing between different clones and varieties, the labels, and that stolons (runners) from each mother plant are trained in one direction to avoid intermixing. (Photo: author.)

Detailed methods are described for doing this for the harmful organisms affecting the crops covered in the scheme guidelines published by EPPO (see Chapter 9). The plant material produced by all these methods must be thoroughly tested and found free of the relevant harmful organisms before it can be accepted as nuclear stock. Nuclear stock must be maintained free from re-infection as far as possible, and may need to be grown under special conditions (Fig. 8.1) at particularly isolated sites or contained in protective structures (Chapter 12).

Components of Quality for Planting Material

During the last half of the 20th century, certification schemes developed enormously in complexity and in the number of crop species covered, largely in response to advances in scientific knowledge and demand from the relevant sectors of agriculture. From relatively simple beginnings, the most sophisticated schemes now incorporate complicated requirements for the maintenance of quality, which fall into three general groups, covering purity, phenotypical characters and health. Each of these components of quality can be controlled in certification schemes to a greater or lesser extent by incorporating

Fig. 8.2. Well grown apple rootstock stoolbeds. When the plants are dormant in late autumn, the stools will be unearthed and the rooted shoots cut off to form new rootstocks. Where rooting is poor, the shoots may need further treatment to promote root formation. (Photo: courtesy of P.J. Reed, CSL.)

appropriate measures, which may include site approval and site freedom from certain pests, soil testing, production and maintenance of nuclear stocks, laboratory testing, pedigree requirements, generation control, germination potential, isolation, agronomy and complex standards for various defects. These defects include the incidence of diseases and invertebrate pests, admixture and adulteration with other varieties, genetic variants, soil, weed seeds or waste matter. The standards are usually known as *tolerances* because they specify to what degree, or how much of, a defect will be tolerated without causing the crop to be downgraded or rejected from the scheme. Schemes may have several different grades or classes, with progressively more relaxed tolerances for successively lower grades. In pedigree schemes the lower grades must be propagated from material of a higher grade or class (in some cases propagation from the same grade is permitted, with a limit on the number of generations at a particular grade). In non-pedigree schemes, grades are allotted on merit, according to inspection and test results. With tree fruit, the operation of certification schemes is complicated by the necessity to cover separately the production of scions and rootstocks (Fig. 8.2) as well as finished trees.

Scheme Procedures

Certification procedures vary between countries and with different schemes. However, typically, procedures start with an application by the grower to enter crops or material into a scheme for inspection. Entries may simply be to a certain scheme (if grades are allotted on assessed merit) or to a particular grade in a scheme. Usually the fee payable must be sent with the application. The application is then checked for completeness and eligibility. In some cases a satisfactory soil test will be a prerequisite to acceptance of the application; if so, applications for and results of soil testing must precede application for entry. Where appropriate, the site or the laboratory where the material is to be grown may then be inspected, and some soil tests may be done at this stage before the site is approved. One or more crop inspections will then be done during the growing period, according to the scheme and pattern of crop growth. Laboratory tests to confirm or identify pests may be necessary. According to the crop and scheme, other inspections may be made after harvest, when further laboratory or growing-on tests may be done to check on the quality achieved. This is often known as the 'post-control' and can serve to check both health and trueness to variety. It is an important feature of many seed potato certification schemes, in which a sample of harvested tubers is tested for virus infection. However, costs do not permit the testing of very large samples (usually no more than 200 tubers), which prevents the use of such tests to verify conformation to very small tolerances of infection. Their main value, therefore, is to identify stocks of poor virus health in areas of high infection risk. Decisions to accept or reject stocks, based on sample size and the results of laboratory tests, are discussed by Lévesque and Eaves (1996).

New, more rapid, reliable and sensitive methods of testing for pathogens in nuclear stock or routine certification continue to be developed (Chapter 10). However, the introduction of novel testing methods to certification systems must be done with care. It is possible that they may reveal the presence of undesirable organisms in existing material where previously none were thought to exist. Unless arrangements are made to permit the temporary marketing of such material (which may still be the best available), it may lead to a serious imbalance of supply and demand and could cause unwarranted losses to producers.

If the crop or material satisfies the scheme and grade conditions, a certificate to this effect will then be issued. In some schemes, labels will also be issued by the certifying authority, while in others labelling is left to the grower. Finally, the crop or material may be listed by the certifying authority in an annual register of certificates issued. This not only provides information for prospective purchasers as to what certified material may be available in any particular season, but also enables them to check the authenticity of certified material advertised for sale. As the physical preparation of such registers is time-consuming, it is an advantage if they can be provided for access on-line at a website.

Schemes for Vegetatively Propagated Crops

Vegetatively propagated crops that are commonly covered by certification or marketing schemes include the following:

- potatoes (*Solanum tuberosum*);
- berry fruits such as strawberry, raspberry, hybrid berries, currants and gooseberries;
- tree fruits such as apple, apricot, avocado, cherry, citrus, peach, pear, plum and their rootstocks;
- grapevines and their rootstocks;
- bulbs such as narcissus, tulip, onion sets;
- hops (*Humulus lupulus*);
- florist's ornamentals such as carnations, chrysanthemums; and
- ornamentals marketed as pot plants.

There are many other food, ornamental and industrial vegetatively propagated crops that would benefit from schemes for the production of healthy stock. Examples include cassava (*Manihot esculenta*), sweet potato (*Ipomoea batatas*), taro (*Colocasia esculenta*) and sugarcane. Schemes for some of these are at the planning or development stage. Certification and marketing schemes are of most use where there is a free and active market in planting material. However, many of these crops are grown primarily either in peasant farming systems or on large commercial plantations. Peasant farmers can seldom afford to purchase high quality planting material unless this is subsidized, and sometimes cannot adhere to the discipline needed to derive maximum benefit from certified material. This may prevent the establishment of certification schemes. The owners of large plantations often feel that they can produce good quality planting material of their own, and may operate on certification scheme principles but without a formal scheme. Official certification schemes for micropropagated material (see below) can sometimes be used successfully to cover planting material of such crops as oil palm, banana and pineapple for supply to plantation growers.

The components of quality for planting material for vegetatively propagated crops, and the ways in which they can be controlled in certification schemes, are reviewed briefly below.

Botanical purity

In modern agriculture it is extremely important for growers to be confident that the planting material they purchase is indeed the variety or type they expect it to be. The crop variety may be specified in contracts for cropping and marketing, and in any case it would be disastrous if a planting stock did not possess the desired characters through error in variety determination or because of major adulteration by mixture with another variety. For example, if a potato variety with poor crisping quality was inadvertently grown instead of that intended, the crop would probably be unsaleable to the crisping factory and, in addition, other consequential losses might ensue. In many countries there are now legal requirements prohibiting the growing of potato varieties not immune to wart disease in the area of wart disease outbreaks, so again, in this case it would be essential for the potato stock planted to be true to its variety and hence its wart-disease-immune character.

Apart from mixture with other varieties, soil and crop debris, botanical purity may also be affected by genetic mutations. Varieties that have been released on to the market are normally genetically stable and should not show variations due to genetic segregation. In the European Union and some other countries these aspects are controlled by a National Listing system, preventing the marketing of unstable or poorly developed varieties (see below). However, even with vegetatively propagated crops, certain somatic mutations occur regularly, and other variants may occur sporadically. With potatoes, mutations producing plants with certain characteristic habits, known as 'bolters' (tall, few-stemmed, robust plants) and 'wildings' (short, many-stemmed, weak plants with a characteristic leaf shape), occur so regularly that tolerances for these are often included in the certification standards. Numerous other undesirable variations occur from time to time. For example, 'June yellows' of strawberries, which results in yellow or white patterns on the leaves and debilitated, small plants, is attributed to genetic faults and can be so serious as to eliminate a variety, as occurred in the UK with the varieties Huxley (in the 1940s) and Cambridge Favourite (in the 1980s). Chimeras, resulting from mutations in the outer layers of cells, produce yellow or pale green variegated foliage in most field crops and in many ornamentals (where they may, or may not, be desirable). It is also possible for mutations to occur in other, less visible characters, such as the time of fruit ripening, but in most cases these are difficult or impossible to detect during field inspections and are usually not possible to control by certification or marketing systems.

With tree fruit species, mutations quite frequently occur in fruit characters, such as skin colour or texture, or in tree habit. Where these are desirable, as with brighter skin colour or more compact tree shape, such mutations are eagerly selected and may be propagated as new 'clonal varieties'. This can create difficulties for the certification authority and its inspectors, as these characters are not often visible at the time of inspection, when young trees of these 'clonal varieties' will be indistinguishable from the 'parent' varieties from which they have been derived. With some species, such as cherries and blackcurrants (*Ribes nigrum*), even normal varieties may be very difficult to distinguish in the young vegetative state. In these cases, therefore, it may have to be accepted that the clonal variety, or even the normal variety, cannot be confirmed during visual inspection. The certification authority will then have to decide whether to give no assurance as to the variety, or to rely almost completely on careful labelling and administrative checks and measures during propagation to avoid mistakes. In the case of 'clonal varieties' a compromise option is to certify them as conforming to the general characteristics of the 'parent' variety and leave it to the propagator to authenticate the particular clone. With tree fruit mother trees and some species that fruit within a year of planting, it may be practicable to allow some fruit to form as a check on possible variation in fruit characters. Alternatively, a sample of the planting material may be grown on to fruit under separate official observation as a check. Where two varieties are vegetatively very similar, a practicable precaution is to prohibit them being grown on the same farm or place of production. In some cases, certain tests (such as the isozyme pattern) can be used to determine the variety in dubious cases, but usually these would be uneconomic or too lengthy to use as routine.

Crop vigour

If planting material is not adequately vigorous, it will usually produce a poor crop, no matter how typical it may be of its variety or how free from infections, although in some cases excessive vegetative vigour can decrease desirable characteristics or commercial yield (as with some virus-free strawberries and rhubarb). Marketing schemes may stipulate the size grading of the progeny material, although certification schemes usually do not, and in many cases this is left to market forces or is covered by trade agreements or practices. However, schemes should contain requirements for good husbandry during the growth of the crop, both to ensure that the crop can be inspected and for the production of good quality planting material.

Health

The pathogens most amenable to control by certification and marketing systems show easily recognizable symptoms of infection in aerial vegetative parts of the host during the growing season, and are carried exclusively in or on the planting material. Only a few pathogens approach this ideal, the best examples probably being wart disease of potatoes (*Synchytrium endobioticum*) and the watermark disease of cricket-bat willow (*Salix alba* var. *coerulea*) caused by the bacterium *Erwinia salicis*. Many diseases, such as chat fruit of apples or plum pox, do not always appear as overt symptoms in the vegetative parts of the host or, as is the case with plum pox, may not be fully systemic, affecting only some parts of the host. Other pathogens of significance in the certification context, such as red core of strawberries and root rot of raspberries (both caused by forms of *Phytophthora fragariae*), may survive as resting organs in the soil, and some have a limited saprophytic life, although their main means of spread is on the planting material or in the soil adhering to it. For some, such as the important viruses of the potato, which cause leafroll and severe mosaic, the pathogens may be transmitted aerially by invertebrate vectors as well as in the planting material.

Provided that a disease is spread mainly by means of the planting material, suitable scheme measures can usually be found to combat the secondary means of spread. This is essential to the success of certification and marketing systems, and measures can be designed to cope with the secondary means of dispersal according to their nature. The pedigree requirement is often used as an additional safeguard for health and for variety authenticity, even where propagation is by conventional means. This is usually implemented by requiring that material entering a scheme has been certified in a certain scheme category or grade (usually a higher category or grade than that at which the progeny will be eligible to be certified).

Measures to control secondary spread by aerial transmission

Most of the pathogens in this category are spread by invertebrate vectors, particularly by aphids. Dispersal can also be by means of wind-blown infected crop trash (as with *Verticillium* wilt of hops) or by pollen (as with many ilarviruses, such as *Prunus necrotic ringspot virus* and *Prune dwarf virus*).

Against all these pathogens the first line of defence is isolation from potential sources of infection. This is based on the observation that the likelihood of transmission from a source of infection to healthy plants declines with increasing distance. In practice, the risk of transmission varies with many factors, such as the mode of transmission and the relationship of the pathogen with any vector, the susceptibility of the host, the efficiency and population of the vector, the size of the infection source and the direction of the prevailing wind. There is also the complication that local spread may be confused with long-distance spread from other sources. A consistent infection gradient is therefore seldom obtained and the risk of spread in any particular situation is difficult to quantify. In practice, the choice of a suitable isolation distance must be somewhat arbitrary and is usually the greatest distance that is practically and economically possible for the average grower.

Removal of infection sources by roguing out infected plants within the crop, or by removing external sources, is another very important line of defence which is effective against all means of spread. The success of this roguing depends on the appearance of recognizable symptoms, their early detection, and on the amount of new infection coming into a stock being less than the amount it is possible to rogue out economically. However, the removal of external sources, such as dumps of crop debris and old roots or tubers remaining from previous crops, is a practical and elementary precaution. The elimination of infection harboured by native vegetation can be much more difficult. For example, the suckers of native *Prunus* species, which are constituents of hedgerows or neighbouring woodland, may harbour infection with *Plum pox virus*. In such cases, successful elimination of infection is seldom successful or economic and it is usually best to find another site for production of the stock to be entered for certification.

Control of vectors is obviously essential in controlling pathogens spread by this means. Before the role of vectors in the transmission of virus diseases was understood, control of aphid-transmitted potato viruses was achieved empirically by growing seed potato crops in areas where they usually escaped infection. Subsequently, it became apparent that the reason for this escape was the generally small populations of potato aphids in these areas and, in particular, the scarcity of viruliferous aphids and the fact that they usually came into the crops late in the season when mature plant resistance and the imminence of harvest combined to inhibit widespread infection and transmission to the daughter tubers. Such favoured areas are often recognized in seed potato certification schemes by defining the areas where certain high grades of seed potatoes can be produced, designating them as Protected Areas or High Grade Regions for the production of high-grade seed, in which low-grade material may not be marketed for seed production or is otherwise controlled. These areas are often in cool upland or windy coastal locations, but in warm climates a similar escape from aphid attack is sometimes possible by production of seed potatoes under irrigation during the dry season, when most of the native vegetation has died back and aphid populations in the surrounding areas are small.

Viruses have various relationships with their vectors and vector aphids can transmit viruses in different ways. The two principal modes of transmission are the *persistent manner* and the *non-persistent manner*. Aphids transmitting viruses in the persistent manner take a considerable length of time (up to about 24 h) both to acquire the virus during feeding from an infected plant and to transmit it during feeding to a healthy plant. Once the virus has been acquired, the aphid remains infective for the remainder of its life. With transmission in the non-persistent manner, however, aphids acquire and transmit the virus rapidly (just a few probes can be sufficient), but may lose the virus and become non-infective if they do not regularly feed on infected plants. This difference in mode of transmission depends on several factors, including the particular transmission characteristics of the virus concerned and whether the aphid has to penetrate to the vascular system of the plant in order to acquire the virus or whether the virus can be acquired from the outer layers of cells into which it first penetrates. It is common to find that several different species of aphids or other invertebrate vectors are able to transmit a particular virus, although it is usual that only one or two of these will be efficient vectors.

Chemical control of aphids is not simple. Aphid populations rapidly develop resistance to most regularly applied aphicides, and may even develop resistance to more than one such chemical. Contact insecticides seldom achieve sufficient cover to give effective control, and a few surviving aphids, as well as being able to transmit virus, can soon multiply to replace those killed. Where the population of aphicide-resistant aphids is negligible, systemic aphicides can be used effectively to control viruses transmitted in the persistent manner, such as *Potato leafroll ilarvirus*. With non-persistent viruses, however, because acquisition and transmission times are short, the vector aphids are unlikely to be killed before transmitting the pathogen. Indeed, in crops treated with certain aphicides, aphids may have a greater tendency to move from plant to plant after the initial probe, sometimes resulting in a greater amount of virus spread than if no aphicide had been applied. More recent techniques of biological and integrated control designed to minimize the use of pesticides (ISPM No. 14; Anon., 2002b) are applicable where consistently small populations of the vector can be maintained. However, they are not appropriate if vector populations increase substantially before control by predators or other biological control agents becomes effective. In some countries, a useful reduction in the spread of non-persistent viruses, including *Potato virus Y*, has been reported after sprays with emulsified oils, but this method has not yet given encouraging results with seed potatoes in the UK.

In seed potato crops a growing season inspection before, or at, flowering time will assess mainly the virus infection carried over from the previous season in the planted tubers (Fig. 8.3). Later inspection may detect new, current season infections, especially in warmer climates. However, infections too late in the current season to produce symptoms by inspection time will not be detected, yet may still penetrate to the daughter tubers. If such late infections threaten to be substantial (which is often a serious possibility in many seed potato producing countries), preventive measures must be taken if assessments of crop health at growing season inspection are not to be invalidated. Besides the control of vectors, therefore, measures often adopted to meet this situation are

Fig. 8.3. Field inspection of a seed potato crop for certification. The plant health inspectors are counting plants that are not true to the variety entered, those which show mild or severe symptoms of virus diseases, and noting attack by any other pests. (Photo: author.)

the destruction of the haulm before viruliferous aphids become numerous, and the post-harvest testing of seed tubers for assessment of virus infection. Such measures are best employed in combination, but they will not be completely effective and not all may be practicable in the prevailing circumstances. Monitoring of aphid populations (see Chapter 10) allows the timely application of precautionary measures.

With some crops, such as hops and strawberries, there is a risk of virus transmission by pollen, that seed may be produced and shed and that arising seedlings may go undetected. Measures to combat this include the deblossoming of mother plants and their isolation from pollen sources. However, this is often difficult to enforce and may also conflict with the need to see a few fruits on plants as a check on variety characters (as mentioned above). Experience in the UK is that the risk of virus transmission by this means in crops entered for certification is small, and that undetected seedlings are seldom a problem.

Control of secondary spread through the soil

Soil-borne diseases that are of significance in certification are mainly caused either by pathogens that are root-inhabiting organisms surviving in the soil, or by pathogens transmitted by soil-inhabiting vectors (such as the nepoviruses transmitted by nematodes and *Potato mop top virus* transmitted by the powdery scab pathogen, *Spongospora subterranea*). As in most agricultural systems, a basic precaution against soil-borne harmful organisms is crop rotation, and schemes should carry an appropriate requirement for this measure. Site approval is another measure for avoiding soil-inhabiting harmful organisms, and is frequently used against potato cyst nematodes in seed potato schemes, where land testing positive for the presence of potato cyst nematodes may not be used for seed potato production. It is possible for transplanted propagating material to carry serious pests for which it is not itself a host in soil attached to the roots. To guard against this, schemes sometimes prudently incorporate requirements prohibiting the growing of crops for certification on land where serious soil-borne pests are known to have occurred, even if these are not pests of the crop to be certified. This precaution is often used, for example, in the case of potato wart disease. If desired, this measure can be reinforced by a requirement for isolation from areas where such pests are known to be prevalent or where they are known to have occurred in the past.

Soil-borne pathogens often cause problems in certification because they commonly do not consistently cause detectable symptoms in aerial parts of the host plant. Crown gall (*Agrobacterium tumefaciens*) of many hosts and leafy gall (*Corynebacterium fascians*) of raspberries are examples. Where such pests are troublesome, it may sometimes be practicable to introduce a requirement for some kind of laboratory or glasshouse pre-planting test to detect them. Otherwise, the random lifting and examination of a few plants at each inspection may be possible as a precaution against their widespread occurrence. However, such precautions must be kept within economic bounds and will only be likely to detect heavy infections or infestations.

Control of secondary spread by contact

Viroids and many viruses spread from plant to plant by means of contact between the organs of neighbouring plants or, through human agency, on clothes, cutting knives or machinery. A prime example of this is *Potato virus X*, but there are many other examples. Clonal selection and multiplication from initially disease-free nuclear stock, in conjunction with tests on field samples and hygienic measures in inspections and agricultural operations, should control this type of pathogen. With potatoes, this means of spread is often encouraged by mechanical methods of haulm destruction, if the crop is not immediately lifted and there is time for transmission to the daughter tubers. The identity of pathogen-tested clones is usually maintained by means of a pedigree requirement (as noted above). This is particularly important in combating pathogens that may not always show detectable symptoms at the time of field inspection.

Rapid Methods of Plant Propagation

Rapid methods of propagation have been developed for many crops, especially for those that are normally vegetatively propagated or are otherwise slow or difficult to propagate. For example, softwood cuttings can be rooted rapidly under mist or fog, bulbs may be 'chipped' or 'twin-scaled', and various forms of tissue culture are widely used for many species. When rapid propagation methods are used, there may be several generations of multiplication before the plants are grown to a suitable size for inspection or to the flowering or fruiting stage. In this situation a check for confirmation of the variety at an early generation of propagation by growing on a sample to flower or fruit is therefore especially important. Some specialized methods of rapid propagation which may have to be considered in certification and marketing schemes are described briefly below.

Tissue culture

For very many plant species tissue culture may be a practicable and useful technique, not only for rapid multiplication, but also to free plants from systemic infections, to facilitate international trade, and to maintain stocks of pathogen-tested material for long periods without risk of re-infection. There are many different variations of tissue culture technique (Zimmerman *et al.*, 1986; Debergh and Zimmerman, 1991), according to the plant species, the variety and the purpose for which it is done. This can include the culture of single cells and the culture of callus tissue composed of undifferentiated cells. However, in certification schemes the use of tissue culture is generally limited to two main techniques. One of these is *meristem culture*, in which just the apical or axilliary meristem tissue is excised and cultured. The other is *micropropagation*, in which the tips of apical or axilliary shoots are excised and cultured. The latter may in fact follow on from the former. The plant part to be used is surface sterilized and the tissue is then excised and cultured on solid (or sometimes liquid) media, which supplies all necessary nutrients and growth-promoting substances, in test tubes or other containers, and is incubated in controlled temperatures under lights. Contaminants or pathogens present in or on the material, which survive the surface sterilization treatment, will often be able to grow on the material in culture or in the medium used for culture, so becoming easily visible and identifying such cultures for discarding. The tissue culture process thus eliminates some pests and automatically reduces the phytosanitary risk considerably. However, some important pests of certain plant species, particularly systemic pathogens such as viruses and viroids, may not be eliminated.

Meristem culture is technically more exacting than micropropagation and success rates are lower. The material excised from the mother plant normally measures no more than about 0.1–0.3 mm in length and consists of the extreme apical dome of dividing cells and perhaps one or two leaf primordia. This meristematic tissue is often free from pathogens even in plants carrying systemic infections. The likelihood that the meristem will be free from pathogens can also be increased by pre-treating the mother plant with carefully controlled heat (for elimination of viruses), cold (for elimination of viroids), or chemicals before tissue is excised. However, the resulting cultured material must be thoroughly tested to verify this. A disadvantage of tissue cultures started from small meristems is that the chance of somatic mutation is increased, and this is a reason why a check on early generations of the resulting progeny for trueness to species and variety is desirable before continuing with mass multiplication of the material. Certain nematode pests (such as *Aphelenchoides* spp. on strawberry) may also survive meristem culture and require special techniques for their elimination (McNamara and Cleia, 1985).

Micropropagation is a form of tissue culture resulting in new plantlets. Usually relatively large excised shoot tips (up to about 5 mm in length) are used as starting material. This has several advantages, provided that elimination of pests is not an objective. The method does not require such sophisticated equipment or such skilled technical personnel as meristem culture, multiplication of the material is usually quicker and it is thus more suitable for routine use. However, the health of the resulting progeny normally corresponds to that of the parent material from which it was started, so it is important to ensure that this parent material is of acceptable health status before it is used for propagation. The chances of somatic mutation are much less than with meristem culture. As with other forms of tissue culture, production of shoot or root growth can be controlled by variation of the growth media and incubation conditions and, where desired, growth can also be slowed in cultures retained for long-term storage. The most vulnerable stage, for both physiological hazards and for infection by pathogens, tends to be at weaning, when the delicate plantlets are transferred from laboratory culture to potting compost or other normal horticultural growing media.

For scheme inspection purposes it may not be possible to detect faults during culture other than obvious contamination, and it can be difficult to arrange for inspections always to be made just before the plants or plantlets are sold and leave the propagator's control. Nevertheless, schemes must keep up with progress in propagation techniques and must accommodate new developments. Where normal inspections cannot be done, the risks of variety muddles or undesirable mutations being multiplied undetected are much increased, and precautionary measures such as growing on samples for inspection, as mentioned above, will need to be taken. Special schemes for tissue-cultured material can be formulated and these can incorporate a pedigree element stipulating the health status and source of the mother plants and requirements for their inspection. To reduce the likelihood of producing undesirable variations, scheme regulations may also stipulate limits on the number of generations in culture or specify the chemical composition or hormone content of culture media. A requirement for meticulous record keeping is also essential. Tissue culture in relation to imports and exports is discussed in Chapter 6.

Bulb propagation

Bulb chipping (Hartmann *et al.*, 1990) is a technique in which the bulb is cut longitudinally into a number of sections (usually 8 or 16, according to bulb size). Each section must retain a piece of the bulb base plate. These sections are then treated with a fungicide and incubated in particular conditions, when small bulblets will form in each bulb section. These can be grown on to form flowering-size bulbs, which may take 2 or 3 years. Twin-scaling is a variant of this process in which the bulb sections are further cut into even smaller sections, each consisting of two bulb scales joined by a portion of the base plate. These are treated, incubated and produce bulblets in a similar way to chips, except that the bulblets are smaller, take longer to reach flowering size, and are much more vulnerable to rotting. Twin-scaling may produce more progeny per bulb in the first generation, but whereas twin-scaling requires considerable technical skill and twin-scale mortality may be high, chipping can be mechanized and results are more consistent, with a greater success rate. These techniques were developed for *Narcissus* bulbs, but can be applied to other species. There are also other specialized techniques for rapid multiplication of bulbs, such as 'scooping' for hyacinth, and many species can be propagated by tissue culture, although for bulbs this tends to have disadvantages similar to those of twin-scaling.

Schemes for Seed-propagated Crops

Except for potatoes, the major food crops of the world and also many important non-food crops are normally propagated by seed. True seeds are usually the product of the sexual process in plants, in which the male gametes in pollen fertilize the ovules in the female ovary to produce an embryo. A notable exception to this occurs in certain groups of plants (not major crops) where seeds are produced by apomyxis, in which the ovules develop into embryos without fertilization. Some species are outbreeding and need to be fertilized by pollen from other plants of the same species. Others are inbreeding and individual plants can be fertilized by their own pollen, while yet others are intermediate between these. Some species possess genetic or physical self-incompatibility systems that render them virtually completely outbreeding, while others possess mechanisms that result in almost complete inbreeding. Many variations fall between these extremes, and most species that can reproduce by seed annually or biennially can be fertilized to some extent by pollen from other individuals of the same species.

Although seed testing for purity and germination potential was started in Germany and Denmark as early as 1869 (Chapter 2), certification schemes for true seed were only developed at about the same time as the early schemes for seed potatoes. Schemes for certification of agricultural seeds were started in the USA in 1915 and the idea was soon taken up by other countries (Parsons *et al.*, 1961; Hewett, 1979). Crops commonly covered by seed certification schemes include the following:

- cereals;
- grain legumes, including beans and peas;

- oil and fibre plants;
- beet;
- vegetables;
- fodder plants, including herbage grasses and legumes.

Seed schemes share the same objectives and many basic features with those for vegetatively propagated crops, and are operated in much the same way, but the emphasis tends to be more on variety identity, purity and germination potential, and less on health.

Botanical purity

Because each seed generation generally involves a certain amount of cross-fertilization between different plants, the genetic constitution of seed-propagated species is often quite heterogeneous. Careful breeding and selection in cultivation can reduce this to the point where varieties breed true from generation to generation but, except for species that are almost exclusively self-pollinating, there is, nevertheless, much greater opportunity for variation than with vegetatively propagated species. Therefore, in schemes for seed-propagated species, much greater attention has to be paid to prevention of inadvertent cross-fertilization and to checking conformation to variety type. Harvest of seed inevitably also gathers in many impurities, including soil, grit and many different kinds of weed seeds, some of which may be very similar to those of the crop to be certified. This necessitates what is sometimes a complicated cleaning and grading process. During this, most (but usually not all) of the impurities are removed, and the process may also include treatment of the seed with a pesticide, coating with a fertilizer, polishing or abrading the seed surface to manage germination potential, drying, grading or a combination of these or other treatments. Handling and moving seed for these processes creates further opportunities for mixture and contamination, against which additional safeguards may be required.

Variety purity is normally determined during field inspection by counting plants or shoots in rows or quadrats. Contamination by weed seeds is normally assessed in the laboratory.

Precautions in the field

As with vegetatively propagated crops, isolation of the seed crop in the field is the principal means of defence against contamination coming from sources outside the crop. However, in this case it is usually unwanted pollen against which the precaution is taken. Pollen can be wind-borne or carried by bees, other insects, or vectors of other kinds. Depending on the mode of conveyance, it can travel great distances but, as with infectious propagules, the risk of receiving unwanted pollen generally declines sharply with distance. The risk also varies greatly with the nature of the incoming pollen and the degree to which the seed crop is outbreeding. If an incompatibility or male sterility factor is operating in the seed crop or potential source of contaminant pollen, the risk may be very small. There is also little risk of viable seed being produced from inter-crossing of diploid and tetraploid varieties. In such cases, therefore, isolation distances can be decreased. Protection can also be provided by removing potential sources of unwanted pollen by mowing off the unwanted flowers before anthesis, or before unwanted seed matures.

As with vegetatively propagated crops, specified isolation distances have to be rather arbitrary and are commonly a compromise, being the greatest distance that is economically comfortable for the average seed grower. Such distances usually vary from 2 to 1000 m, according to the crop, the assessed risk of unwanted cross-pollination and the grade of certification. Because the risk of cross-fertilization seldom declines to zero, great care must be taken in situations where even slight contamination with unwanted pollen must be avoided. Sources of contamination from within the crop must also be eliminated, and the removal of plants that are of a different variety or are not true to the characters of the variety being produced is an important operation in the cultivation of seed crops. This can be particularly important in the production of hybrid seed where male-sterile lines rely on specific pollinators for fertilization.

In recent years the need for preventing unwanted cross-fertilisation has acquired a high profile in the production of seed of varieties resulting from genetic modification (GM) of some kind. There have been several notable cases where genes from GM varieties have been identified contaminating non-GM varieties (Brookes, 1998). Despite wide isolation requirements, in such cases it may be necessary to avoid GM and non-GM varieties of the same or closely related species being grown on the same holding, or being handled by the same seed-processing plant. There is also the risk that cross-fertilization with related wild species may allow unwanted GM genes to spread into the native flora, with potentially serious consequences. For example, native weeds may acquire herbicide resistance, or species may acquire resistance to native insects that are a food source for wildlife. If this is a risk, there may be no alternative but to require production of the GM seed within a securely protected environment, or in foreign areas where wild relatives do not occur. For example, there should be little risk of contamination between maize and the native flora of Europe, where it has no close relatives, although there might be a risk to other maize crops.

In many seed schemes, especially at the higher grades, it is usual for the certifying authority to grow small reference plots from seed samples taken from the parental seed bulk before sowing. Reference plots may also be grown from a proportion of the final generation of seed (the lowest certification grade). These plots provide valuable information for seed crop inspectors and may serve to confirm or refute faults found during field inspections. They also provide opportunity for additional checks in case of disputes, and those from the lowest certification grade provide information on the quality of the seed leaving the certification system.

Most seed schemes contain conditions for previous cropping and length of rotation on the land where the seed crop is to be grown. This guards against the appearance of 'volunteer' plants deriving from dormant seed shed from a previous crop, and is particularly helpful in preventing contamination with weed relatives of the seed crop, such as weed beet in beet, or wild oats in cereals. Proper use of pre- or post-emergence herbicides also has a part to play here.

Vigour and physiological characters

Seed crops must be sufficiently vigorous to allow the variety characteristics to be fully expressed and to give a reasonable yield of seed, so many schemes contain requirements for good husbandry. As with vegetatively propagated crops, the seed crop must be sufficiently well grown and free from weeds to be inspectable at the proper time to distinguish the variety characteristics, which may otherwise be masked or poorly displayed. Seed processing and treatments of various kinds can improve the vigour of the marketed seed by removing weed seeds and poorly developed seeds of the seed crop. Treatments to improve germination or to improve the quality of the seed for sowing can also be applied. For example, the seed may be scarified, either to improve its ability to absorb water and thus enhance germination or, in the case of beet, to eliminate some of the seeds in the seed clusters so that generally only one seedling will result.

Health

As already noted, in general health tends to receive less emphasis in seed schemes than in those for vegetatively propagated crops. Although there are a large number of seed-borne pathogens, they tend to be more easily controllable with pesticides and do not accumulate and cause degeneration of the crop in the same way as with vegetatively propagated crops. A comprehensive guide to important seed-borne pathogens and seed health testing is provided by Hutchins and Reeves (1997). Many viruses and other systemic pathogens do not enter the seed and so are not seed-borne. However, some non-systemic pathogens contaminate the surface of the seed and are seed-borne in this way. Many of these can be controlled easily by some form of seed treatment, such as coating with a pesticide or disinfection with chemicals, and some schemes contain requirements for this, which may be conditional on the result of a laboratory test for the pathogen concerned. Some pathogens do enter the seed and may be present within the seed coat. For example, the mycelium of loose smut (*Ustilago nuda*) infects the embryo of barley seed. A laboratory test for this, involving the microscopical examination of embryos extracted from a seed sample, can give a good indication of the prevalence of loose smut in the seed tested. Where the seed is infected internally, external seed treatment will not usually control the pathogen unless the material penetrates and acts systemically. Such pathogens are best controlled in the seed crop before harvest, to prevent them from entering the seed.

Seed-borne diseases are commonly caused by fungal pathogens, such as ergot (*Claviceps* spp., particularly *C. purpurea*), smuts and bunts (*Ustilago* and *Tilletia* species) in cereals and grasses, and *Ascochyta*, *Fusarium*, *Phoma* and *Septoria* species in many crops, including cereals. However, some serious bacteria and virus pathogens are seed-borne, such as *Pseudomonas syringae* pathovars in peas and beans, and *Lettuce mosaic virus* in lettuce. These are difficult or impossible to control by seed treatment and must be controlled in the field.

Control measures

Many of the measures mentioned in relation to vegetatively propagated crops, or for prevention of contamination by pollen, are also beneficial in controlling pathogens in seed crops. Disease escape by means of isolation, or growing the seed crop where risk of attack by seed-borne pathogens is small, is therefore important. Site approval, and requirements for length of crop rotation, previous cropping and roguing of plants that may be sources of infection can all be helpful. Control of vectors is relatively less important, as most seed-borne pathogens are also dispersed by wind or rain splash and can usually be well controlled by application of suitable pesticides, notable exceptions being bacteria and *Fusarium* spp.

Seed treatments form a branch of pesticide technology that cannot be covered in detail here, but these are very important in promoting healthy seed-sown crops. Current technology and developments in this area are covered by Biddle (2001). As with pesticides applied to crops in the field, active materials in seed treatments may act by contact or they may penetrate and act systemically within the seed to confer protection from attack or to inactivate pathogens already present within the seed. Their action thus occurs within the interior or on the surface of the seed, and in the immediate surroundings of the seed and the germinating seedling in the soil after sowing. Their effects do not persist for very long after germination. Chemical seed treatments may be applied in liquid, powder or slurry form, or they can be incorporated into coatings covering the seed to form a pellet. In recent years pelleting technology has developed rapidly and pelleted seed has become increasingly commonly used. The pellet coating may incorporate a fertilizer as well as a pesticide, and may also confer other benefits, such as facilitating the precision sowing of the seed.

Laboratory tests

Laboratory tests for both physiological quality and health are extremely important in seed certification. Such tests, done on representative samples, can identify seed stocks that should not be certified on account of poor quality in these respects (Hutchins and Reeves, 1997). The seed samples must be drawn in such a way that the sample adequately represents the seed bulk being tested, and the methods for drawing such samples are often specified in considerable detail by international or national regulations. Samples may be drawn by official personnel or by others officially licensed and trained to do so. Conditions for seed schemes normally contain standards for germination potential and a maximum tolerance for contamination by seeds of weeds and other crops, visible variants of the crop species, and other contaminants such as fungal sclerotia, soil and crop debris. According to the crop species and the requirements of the certification scheme, or the regulations of the importing country, tests may also be done for specific seed-borne pathogens and invertebrate pests. Every seed-certifying authority requires access to a seed testing laboratory capable of reliably assessing seeds for these criteria. Seed health tests may use serological and molecular-based methods (Hutchins and Reeves, 1997) as well as more traditional tests and growing on of seedlings (see Chapter 10).

An OSTS is a seed testing laboratory officially designated and approved for this purpose by the seed certifying authority of the country concerned, and normally one such laboratory is sufficient for each country. However, to minimize costs and to cope with heavy workloads, a system is often operated whereby other laboratories are licensed to do some of the seed testing. For international recognition, OSTS laboratories must be affiliated to ISTA, which internationally coordinates seed testing and the issue of international certificates for seeds. Methods for sampling seed lots and laboratory procedures for determining germination potential, seed purity and infection by seed-borne pathogens are developed or evaluated by the ISTA Plant Disease Committee and published in the form of Working Sheets. These are incorporated in the ISTA *Handbook of Seed Health Testing*, which also includes an annotated list of seed-borne diseases. Similar guidance on methods is also published by the Association of Official Seed Analysts in North America. The ISTA provides training in the form of courses and workshops, which aim to improve expertise and harmonize methods in seed testing and so promote consistency and comparable results.

Potatoes From True Seed

In recent years, considerable research has been done on growing *Solanum tuberosum* potatoes from true seed. As yet, it does not seem that such seed has become the subject of a national certification scheme, but the International Potato Center (Centro Internacional de la Papa) at Lima, Peru, distributes much true seed of potatoes worldwide and devotes considerable attention to its quality. Because potatoes are genetically normally very heterogeneous, potato plants and their tubers grown from true seed have usually been too variable for commercial purposes. However, true seed of some varieties gives less variable plants (which may be acceptable for home cultivation and use) and improvements in uniformity have been made by selection and breeding. True seed for potato crops is especially useful in frost-free climates, or where the growing season is sufficiently long to permit planting out the delicate seedlings after all risk of frost has passed. The advantages include automatic freedom from non-seed-borne pathogens (most of the important potato viruses) and the ease and cheapness of transportation and storage. Disadvantages may include the need for transplanting, more intensive husbandry, and great susceptibility to virus infection at the seedling stage. Potato crops from true seed have done well in peasant farming systems in upland tropical areas. When better true seed varieties become available they may be more widely adopted commercially, and true seed of potatoes may feature in seed certification schemes.

Inspections

Visual inspection is the principal means used in certification and marketing schemes to detect defects, and inspectors require regular and thorough training for this part of their work (see Chapter 5). Most botanical variants and mixtures in a crop are visible as a lack of uniformity to the trained eye of experienced inspectors making inspections during the growing period. Mild symptoms of virus infections can be difficult to distinguish, and training for inspectors on recognition of such symptoms may be important. Use of kits for the identification of viruses and other pathogens in the field (Chapter 10) can provide back-up for visual inspections. Plants must also be in a satisfactory state for inspection, being not only at a suitable stage of growth to reveal defects, but also reasonably free from weeds, reasonably well-grown, and without atypical growth due to the application of agricultural chemicals that affect the growth habit.

Voluntary and Compulsory Schemes

The question as to whether certification and marketing schemes should be voluntary or compulsory can generate much debate. When such schemes were first introduced, they were usually voluntary, but when their effects on improvement of the quality of planting material and the consequent increase in crop yields were seen to be so great, schemes in some countries were made compulsory by requiring that all material traded had to be certified. Clearly, both voluntary and compulsory schemes each have certain advantages. However, it should be kept in mind that the nature of a scheme should be appropriate not only for the crop and industry it serves, but also for the social structure in which it operates. As mentioned earlier, many crops that would be appropriate subjects are not covered by official schemes because of the farming systems in which they are grown.

Compulsory schemes are desirable for improvement of the quality of propagation and planting material where necessary measures would otherwise not be applied. They should raise the general degree of confidence in the national quality and uniformity of the planting material produced. They also give the certification authority greater control over the quality of the crop planting material and may make it easier to combat problems for which widespread and long-term action is required. They are also advantageous where the owners of plant variety rights or patents (see Chapter 9) are reluctant to enter varieties that are in demand. Additionally, compulsory schemes result in a steady, predictable and more substantial income to the certifying authority from the fees charged. This not only permits the authority to support related research to improve the schemes and, where appropriate, to support promotional activities, but economies of scale also allow a reduced level of fees to be charged. However, compulsory schemes require moderately relaxed standards for the lowest grade of certification in order to gain general acceptance that failed material should be unmarketable.

Voluntary schemes retain the element of choice. The propagator may consider that the quality of material produced is already sufficiently good for its purpose or to satisfy customer demand, and he or she may find it preferable and more economic to trade on an established reputation rather than enter an official scheme. Provided good quality planting material is available to the industry and this is properly labelled and publicized, a voluntary scheme allows market forces and the *caveat emptor* factor to operate for uncertified material. There is also the possibility that, with voluntary schemes, material that has been refused certification for some defect, such as slightly exceeding the tolerance for botanical purity, could still be marketed. This may sometimes facilitate more rigorous adherence to stringent scheme conditions and eases the decision of the inspector in borderline or hardship cases.

The benefit of all certification schemes will vary with the degree to which the health and other desirable characters of the propagation material can be controlled by the certification process. For pests that are amenable to control by this means, such schemes can provide good control and often at a relatively small cost. However, although cost:benefit ratios can be very favourable (e.g. Ebbels, 1988), growers may not always be willing or able to meet the full costs of running a scheme. As mentioned in Chapter 6, compulsory certification schemes can be used for control of pests that fall into the category of regulated non-quarantine pests (ISPM No. 16; Anon., 2002c).

Most seed potato producing countries now require all marketed seed potatoes to have been officially certified, and within the EU seed potato certification has been compulsory since the adoption of Council Directive 66/403/EEC in 1966. With other crops, some countries have gone further along this road than others, Denmark and The Netherlands being notable as pioneers of compulsory schemes before the European Union introduced legislation in this area. Statutory standards for the health of many types of planting material were introduced in Denmark in 1982 and for tree fruit material in Germany in 1986. Statutory health standards for bulb crops are of long standing in The Netherlands, where the number of species subject to compulsory certification increased steadily from the 1930s onwards. All marketed propagation and planting material of the major European fruit species and many species of bulbs and glasshouse ornamentals, as well as seed potatoes and seed-propagated field and vegetable crops, are now subject to obligatory certification in The Netherlands through the NAK, BKD (see Chapter 2) or Naktuinbouw organizations (Anon., 2000). The certification authorities of most countries are within the government administration responsible for agriculture (often within the national plant protection organization), or are government bodies of some kind. However, in some countries schemes are operated by universities, agricultural institutes, trade organizations or other non-governmental organizations, which may combine representation from several interested parties, including government.

Keeping Schemes up to Date

Almost every branch of applied biology has some bearing on crop certification, and developments in these areas of science are often followed by developments in propagation methods or in disease detection techniques. This often necessitates revision of scheme conditions to ensure that they are compatible with the latest practices and incorporate safeguards that are based on the latest techniques. Schemes with out-of-date conditions create problems, not only for growers, who may be reluctant to enter their crops, but also for the administrators of the schemes, because of diminished revenue and an increase in queries and complaints. However, schemes tend to become more complicated as rules and regulations are devised or modified to meet new or changing situations. In this there is a danger that schemes may become too complicated or sophisticated for the agricultural systems they serve. It is pointless to apply regulations designed for the improvement of already high quality material in circumstances where the basic requirements are not yet being met. Similarly, tolerances for defects should be set at a challenging, yet achievable level. Unrealistically tight tolerances are counter-productive and may lead to evasion by false returns or stock switching.

It is good practice to review, and if necessary revise, scheme regulations annually, or at least at regular intervals. No schemes can survive unless they have the support of the agricultural or horticultural industry they serve, and so it is highly desirable that all parts of the industry should have input to these reviews and have opportunity to propose, comment on and agree any changes. The reviews should therefore involve representatives of the growers, dealers, scheme administrative, inspection and scientific authorities, and any trade or professional bodies involved with the trade sector concerned.

International Certification and Marketing Schemes   9

Introduction

Several different international organizations have developed schemes to improve the quality and health of material used for propagating agricultural, horticultural or forestry crops. The primary objective of many of these is to encourage international trade, but some (particularly those developed by EPPO) are designed specifically to improve plant health. Brief descriptions of these schemes are given below.

Seed Potato Standard of the Economic Commission for Europe

The United Nations Economic Commission for Europe (ECE) is concerned to encourage economic cooperation among its member states. As part of this objective in the agricultural sphere, it established in 1949 a working party, which developed into the current Working Party on Standardization of Perishable Produce and Quality Development. Within the Geneva Protocol on Standardization of Fruits and Vegetables, the working party formulated and adopted a standard for seed potatoes moving in international trade between and to ECE member countries (including Canada and the USA), which was published in 1961. The Standard has passed through five revisions to date and is intended to be applied by the exporting country. The current text can be found on the ECE website: http://www.unece.org

The ECE Standard on Seed Potatoes greatly influenced the standards and conditions for seed potato production and marketing, which were developed later by the European Economic Community (EEC) and embodied in Council Directive 66/403/EEC (now consolidated in and superseded by Council Directive 2002/56/EC of 13 June 2002). The rules and standards for seed potatoes required by the ECE and the European Union remain very similar. However, both systems are subject to amendment at quite frequent intervals and the current texts must be consulted to determine current points of difference.

European and Mediterranean Plant Protection Organization Schemes

EPPO was the first international body to attempt to formulate certification procedures for crops other than potatoes. The EPPO Working Party on Certification for Virus-tested Fruit Trees, Scions and Rootstocks was inaugurated in 1970. Its objectives were to prepare scheme guidelines or 'blueprints', based on sound and scientifically valid procedures, which could be recommended to Member Countries for adoption when setting up or revising their domestic fruit tree certification schemes. These guidelines included selection, growth and maintenance of the candidate material, means of propagation, relevant pests, testing methods, the most suitable indicator test plants and recommended standards. Meetings were held in 1970, 1976, 1980, 1986 (when the working party became a panel reporting to the EPPO Working Party on Phytosanitary Regulations; see Chapter 3), 1988 and 1989. By this time the work of the panel had expanded to include other fruit species. Accordingly, it was agreed appropriate to change its name to Panel on Certification of Pathogen-tested Fruit Crops, as which it has met more frequently thereafter. A scheme for temperate tree fruits was published in several parts during 1991 and 1992, including important definitions of terms such as 'virus free' and 'virus-tested', together with lists of the pathogens from which each category of stock should be free, how the necessary testing should be done, and standards to be satisfied by enterprises propagating certified fruit trees. Other schemes were published in later years and, at the time of writing, the list of EPPO fruit schemes is as given in Table 9.1. The original scheme for tree fruits has been revised and published as three separate schemes (Table 9.1).

In 1986 the EPPO Panel for Certification of Ornamental Plants was formed, with similar status and objectives to its sister panel on fruit crops. It has met frequently since then (more recently under the title of Panel on Certification of Pathogen-tested Ornamentals) and has published certification schemes for a number of ornamental species, as shown in Table 9.1. More are planned. Because it was recognized that the health status of many ornamentals was much less well known than those for fruit, it was felt that there was a need for lower-grade schemes based mainly on visual inspection and which did not envisage a pedigree of descent from fully pathogen-tested nuclear stock. Work was therefore started on a series of schemes in a second category to cover such material and to be known as Classification Schemes. In general

Table 9.1. European and Mediterranean Plant Protection Organization (EPPO) schemes for the production of healthy plants for planting.

EPPO Standard	Scheme or standard	Published	EPPO Bulletin reference
	EPPO Standards and definitions	2002	**32**, 49–53
PM 4/7(2)	Nursery requirements	2001	**31**, 441–444
	Certification schemes		
PM 4/1(1)	Fruit trees and rootstocks, Part I	1991	**21**, 267–277
	Parts II–IV	1992	**22**, 255–283
	[all superseded by PM 4/27(1), 4/29(1) and 4/30(1)]		
PM 4/2(2)	Carnation	2002	**32**, 55–66
PM 4/3(3)	Pelargonium	2002	**32**, 67–78
PM 4/4(2)	Lily	2002	**32**, 79–90
PM 4/5(2)	Narcissus	2002	**32**, 91–104
PM 4/6(2)	Chrysanthemum	2002	**32**, 105–114
PM 4/8(1)	Grapevine varieties and rootstocks	1994	**24**, 347–367
PM 4/9(1)	*Ribes* (currants and gooseberries)	1994	**24**, 857–864
PM 4/10(1)	*Rubus* (raspberries, blackberries, hybrid berries)	1994	**24**, 865–873
PM 4/11(1)	Strawberry	1994	**24**, 875–889
PM 4/12(1)	Citrus trees and rootstocks	1995	**25**, 737–755
PM 4/16(1)	Hop (*Humulus lupulus*)	1997	**27**, 175–183
PM 4/17(1)	Olive trees and rootstocks	1997	**27**, 185–193
PM 4/18(1)	*Vaccinium* spp.	1997	**27**, 195–204
PM 4/19(2)	*Begonia* spp.	2002	**32**, 135–146
PM 4/20(2)	*Impatiens* New Guinea hybrids	2002	**32**, 147–158
PM 4/21(2)	Rose	2002	**32**, 159–178
PM 4/25(2)	*Kalanchoe* spp.	2002	**32**, 199–210
PM 4/26(2)	Petunia	2002	**32**, 211–221
PM 4/27(1)	Apple, pear and quince	1999	**29**, 239–252
PM 4/28(1)	Seed potatoes	1999	**29**, 253–268
PM 4/29(1)	Cherry	2001	**31**, 447–461
PM 4/30(1)	Almond, apricot, peach and plum	2001	**31**, 463–478
	Classification schemes		
PM 4/13(2)	Tulip	2002	**32**, 115–122
PM 4/14(2)	Crocus	2002	**32**, 123–128
PM 4/15(2)	Bulbous iris	2002	**32**, 129–134
PM 4/22(2)	Freesia	2002	**32**, 179–184
PM 4/23(2)	Hyacinth	2002	**32**, 185–190
PM 4/24(2)	Narcissus	2002	**32**, 191–198

terminology, these may be called 'marketing' schemes, but much of the information they contain is similar to that of the certification schemes. Several such schemes were completed and revisions have been published, as shown in Table 9.1. In 2000 the schemes listed were collected together and published by EPPO as a booklet. In addition to the panels on fruit and ornamentals, an EPPO Panel on Certification of Seed Potatoes was formed, which operated over several years for the preparation of a scheme for seed potatoes, which was published in 1999 (Table 9.1).

EU Schemes and Related Regimes

Seed potato certification

As mentioned above, in 1966 the EEC agreed a Council Directive on marketing standards for seed potatoes, which received many amendments over the succeeding years. This has now been consolidated in a new Council Directive, 2002/56/EC. Seed tubers that do not conform to the requirements of the Directive may not legally be marketed within the EU. The Directive not only sets minimum quality standards for seed potato tubers to be marketed, but also lays down minimum standards that are required to be met by the daughter crops grown from such tubers. However, the means by which these standards are to be achieved are not specified, and it is left to each Member State to adopt regulations and production methods suited to its particular agricultural and legislative systems. Member States may also adopt more stringent standards for seed potatoes produced within their own territory.

Under this legislation, seed tubers for marketing are divided into two categories or classes: Basic Seed, which is intended mainly for use in producing certified seed potatoes (i.e. another seed tuber crop), and Certified Seed, which is intended mainly for use in producing potatoes other than seed potatoes (i.e. ware crops for consumption or industrial use). To avoid confusion when referring to certified seed potatoes of Basic Class, the scheme procedure was referred to as 'classification' (i.e. placing seed tubers in one or other of the two classes). However, the introduction by EPPO of 'Classification Schemes' introduced a further complication, and one must now rely on the context to make the meaning clear. Member States may choose to subdivide the two classes into grades with various standards, which must nevertheless conform to the class standards specified. There is also provision permitting the marketing of seed potatoes of Pre-basic generations, intended mainly for the production of Basic Seed. In addition to these classes, a series of Community Grades has been agreed (Commission Directive 93/17/EEC), namely EC1, EC2 and EC3, which may or may not correspond more or less to the grades established by the individual Member States. Seed potatoes conforming to one of these Community Grades may be marketed as such within EU-designated High Grade Regions as well as elsewhere throughout the EU.

The minimum quality standards specified in Council Directive 2002/56/EC are more stringent for Basic than for Certified classes of seed potatoes. For the seed tubers to be marketed, the standards cover tolerances for specified wet and dry rots, common scab (*Streptomyces scabies*), blemishes and misshapes, size, and soil and extraneous matter. All tubers are required to be free from wart disease (*Synchytrium endobioticum*), ring rot (*Clavibacter michiganensis* ssp. *sepedonicus*), brown rot (*Ralstonia solanacearum*) and the production site must be found free from potato cyst nematodes (*Globodera rostochiensis* and *G. pallida*). Requirements for packaging, sealing and labelling are also specified. The package of seed potatoes must contain an Official Document giving at least the same information as the label, unless the label is printed indelibly on the exterior surface or consists of tear-resistant material. The use of sprout inhibitors on seed tubers is prohibited, and any chemical treatments applied to the tubers must be stated on the label and Official Document.

The Directive also specifies minimum quality standards for the direct progeny of the marketed tubers (i.e. the crop to which they give rise), covering the incidence of blackleg (*Erwinia carotovora* sspp.), plants not typical of the variety being grown, and plants with symptoms of viruses occurring in Europe. The only way to be certain of the incidence of these defects in the progeny crop is to plant the seed tubers or to test a large number in the laboratory, which would be very costly. Conformation to the standards is therefore satisfied by a combination of measures designed to prevent the spread of infection and a tolerance for virus incidence in the seed tuber crop, which is sufficiently small so that in most years the amount of virus multiplication does not result in more than the specified tolerance in the progeny. Where the efficacy of this arrangement would frequently be in doubt, the incidence of virus in the seed tubers to be marketed can be checked (within confidence limits depending on the size of the sample) by testing a sample of the tubers in the laboratory. Many countries do this as a routine part of the certification process.

Administration and amendment of the Directive is governed by the Standing Committee on Agricultural, Horticultural and Forestry Seeds and Plants (see Chapter 4). There are provisions for emergency measures, for example, if there is a shortage of seed tubers of the required standard of a particularly desired variety, or if seed potatoes from one Member State are perceived to pose a health threat to another Member State. There are also provisions that enable seed potatoes produced to equivalent standards in a third country to be recognized as equivalent to those produced within the EU. These are eligible to be marketed in the EU under the same conditions as those produced within the EU. To check quality and to ensure the uniform application of standards throughout the EU, a system of annual comparative trials was established. For these trials samples are drawn from the various classes of classified seed potatoes produced in Member States and third countries with recognized equivalence. These samples are grown alongside each other at a single location and are assessed for their conformation to the required standards by a Committee of Experts drawn from the participating Member States.

Marketing schemes for seed of agricultural and vegetable crops

The EEC introduced legislation for the control and certification of agricultural field crop seeds at about the same time as for seed potatoes, in 1966. Schemes and regulations covering seed of beet, cereals and fodder plants were later followed by schemes for seed of oil and fibre plants and for vegetable seed, as summarized in Table 9.2.

This legislation prohibits the marketing of seed of the crops covered unless it has been officially certified and the variety appears in the appropriate National List or the Common Catalogue (see below). Seed is certified in various categories, depending on the crop species, which may be Breeder's Seed, Pre-basic Seed, Basic Seed or Certified Seed of first, second or third generation. There are also the lower categories of Standard Seed and Commercial Seed, which can be used for vegetables and for certain agricultural crops, respectively. Seed merchants, packers and processors are required to maintain proper records of seed transactions, testing and treatments applied. Seed and seed crop inspection, sampling and testing must be done officially or under official control (for which a licensing system is under trial).

These seed certification schemes set standards for variety identity and purity, analytical purity, germination capacity, content of weed and other crop seeds, and for freedom from certain seed-borne diseases. Compliance with these standards is checked by the testing of official

Table 9.2. European Union Marketing Directives for seed of agricultural crops.

Council Directive	Covering (as amended)
66/400/EEC	Marketing of beet seed
66/401/EEC	Marketing of fodder plant seed
66/402/EEC	Marketing of cereal seed
69/208/EEC	Marketing of oil and fibre plant seed
70/458/EEC	Marketing of vegetable seed

seed samples at an official seed testing station or licensed seed testing station. There are provisions for packaging, sealing and labelling, including the information to be stated on the label, for marketing of seed mixtures and for the production and certification of hybrid seed. Again, the means by which these standards are to be achieved are not specified but are left to individual national governments. As with other EU schemes, there is provision for granting equivalence to seed produced to equivalent standards and under comparable conditions in third countries. This is facilitated where these are members of the Organization for Economic Cooperation and Development (OECD) or are participants in the OECD seed schemes.

Other marketing schemes

Common quality standards for marketing of nursery stock, including fruit plants, were introduced by the EEC in 1968, but these concentrated on size and presentation, making little mention of health. Later, an *ad hoc* Committee of Experts for the Elaboration of Certification Schemes for Fruit Tree Reproductive Material was convened and met in 1986 and 1987, without any progress. However, because various provisions for statutory standards in some Member States were seen as a potential barrier to trade, the introduction of more comprehensive common standards as part of the Single Market initiative was felt to be desirable and would also improve standards where no statutory standards existed. In 1989 the EC Commission made proposals that eventually resulted in Decisions and Directives establishing marketing regimes for: (i) ornamental plant propagating material and ornamental plants; (ii) vegetable propagating material and vegetable plants (excluding seeds); and (iii) fruit plant propagating material and fruit plants. Although there was some controversy concerning the necessity for such measures, especially for ornamentals, legislation in the form of Council Directives was eventually agreed for each area by 1992. Implementing Commission Directives were agreed in 1993 in time for the introduction of the Single Market plant health regime (see Chapter 4). However, as this legislation began to be implemented in practice, the feeling that the requirements for ornamentals were over-prescriptive became stronger. The list of species and varieties to be covered had been drawn up with little discussion, and several of the items had no serious health or other problems that could be controlled by such a scheme. As time went on, the opinion that many items were included for no good reason gained wider acceptance and, with a view to revision of the legislation, the original Council Directive, 91/682/EEC, was included in the Commission programme for Simpler Legislation for the Internal Market, which was launched in 1996. During 1998 and 1999 the legislation

Table 9.3. European Union Marketing Directives for ornamentals, vegetable plants and fruit.

Council Directives	Crops	Commission Directives	Content
98/56/EC	Ornamentals		Provisions for marketing
		93/49/EEC	Conditions to be met
		1999/67/EC	Amendments
		1999/68/EC	Provisions for variety lists
92/33/EEC	Vegetable plants		Provisions for marketing
		93/61/EEC	Conditions to be met
		93/62/EEC	Measures for supervision
92/34/EEC	Fruit		Provisions for marketing
		93/48/EEC	Conditions to be met
		93/64/EEC	Measures for supervision
		93/79/EEC	Provisions for variety lists

covering the marketing of ornamentals was revised. Agreement was reached on a new Council Directive and implementing Commission Directives, as shown in Table 9.3. In general, these had the effect of making the provisions covering the marketing of ornamentals more logical and less onerous for producers and the certifying authorities.

For ornamentals the legislation now covers all species except where material is to be exported to third countries, where the products are not intended for ornamental purposes, and the seeds of certain species are exempted. Requirements are based on visual characters and merely require material to be of 'satisfactory' quality and 'substantially' free of harmful organisms or defects that would impair its usefulness. Rather more specific conditions are specified for *Citrus* material and for flower bulbs, and provide for the formulation of additional conditions for genera or species where there is a demonstrated need for this.

The Council Directives for material of vegetable and fruit species have many points in common and provide a framework for the implementing conditions contained in the Commission Directives. The common points include requirements for cultivation, botanical purity, trueness to variety, the making of checks and the maintenance of records (which are not required for those dealing entirely with retail or the local market), the accreditation of suppliers and testing laboratories, the arrangements for assuring trueness to variety (varieties having to be either protected under plant variety rights, as described below, or described in suppliers' lists), packaging, sealing and labelling, official inspections and sampling. There are also provisions for the granting of equivalence to material produced in third countries and for the conduct of trials or tests to verify compliance with the required standards.

For vegetables, the legislation covers only vegetative material of those vegetables covered by the vegetable seeds marketing Directive (70/458/EEC) with the addition of *Allium cepa* var. *ascallonicum* (shallots), *A. fistulosum* (Japanese bunching onions, but listed as chives), *A. sativum* (garlic), *Cynara scolymus* (globe artichoke) and *Rheum* (rhubarb). For the preservation of older vegetable varieties, exemptions are permitted for the purpose of preserving genetic diversity. Again, many vegetables in the list have no serious health problems that could be controlled by such a scheme, and there appears to be little reason for their inclusion.

As many, but not all, Member States had developed their own certification schemes for fruit propagating and planting material, and the benefits from certification were more obvious, there was good reason for developing a Community scheme for this sector. However, although the legislation incorporates many of the basic principles from schemes developed by individual Member States, and indeed relies heavily on the concepts and requirements developed by EPPO, this scheme was developed in haste and does not confer the benefits that it could yield. There is provision for both virus-free and virus-tested material (following the EPPO definitions), and for two marketing categories. There is a lower category of compulsory minimum standard, designated CAC (*Conformitas Agraria Communitatis*), for which standards and procedures are quite relaxed, and a voluntary higher category with more stringent requirements. This is divided into Pre-basic, Basic and Certified grades, which must derive from tested Nuclear Stock and for which the pedigree of material marketed is specified and must be recorded. As for vegetables, there are provisions designed to preserve genetic diversity and arrangements to permit the marketing of old fruit varieties that are not protected under plant variety rights. The scheme covers all the common fruits and nuts found in temperate and Mediterranean climates including, where appropriate, seeds and rootstocks. However, there appear to be no good health reasons for including some subjects (for example, *Pistacia vera*, the pistachio nut).

National Lists and Common Catalogues of varieties

Although not part of the EU plant health regime, the National Listing system is an integral part of the EU certification and marketing systems for seed and propagation material and must be taken into consideration when such systems are operated. Within the EU, each Member State is required to produce National Lists of varieties that have been officially accepted for certification and marketing in its territory (except for fruit and ornamental varieties, to which this does not apply). Marketing of unlisted varieties is normally prohibited. There is a requirement for official trials and tests to determine whether a variety qualifies for inclusion on the National List, and the variety must have an official description. For acceptance on to a National List, a variety must be Distinct, Uniform and Stable, and for most crops (not vegetables) it must also have Value for Cultivation and Use. To qualify under these terms, a variety must be distinct in at least one character from any other variety known in the EU; the plants comprising a population of the variety must conform to the variety description and its distinct characters; it must remain true to the description of its original characteristics through successive generations; and it must show a clear improvement over varieties already on the National List or Common Catalogue, or have some special characteristic that compensates for inferior performance in some respect. When a variety has been accepted on to a National List, it is also entered on the EU Common Catalogue of varieties, and is then eligible to be marketed throughout the EU. Each variety entered for National Listing is required to have a maintainer undertaking to maintain the variety true to type and keep it available. The National List also contains the names and addresses of maintainers of all listed varieties, and if no one wishes to maintain a variety any longer, it must be deleted from the List. Council Directives 70/457/EEC and 70/458/EEC establish this system for agricultural crops and for vegetable species, respectively.

Plant variety rights

Plant breeders very often invest substantial resources of time and money in the production of new plant varieties. If, when these are released on to the market, anybody can multiply and market them for their own gain, the breeders are at a disadvantage, in the same way as authors without copyright for books. The breeders need to recoup their investment before making a profit and, before the introduction of plant variety rights legislation, there was usually only a short time for them to do this before the variety was being produced by all competitors in the market, especially with good new varieties. This was not important when most major plant breeding programmes were funded by governments but, with the increase in private commercial plant breeding, it was increasingly felt to be unsatisfactory. Protection for the breeders in the form of a Plant Breeder's Right for new varieties was therefore introduced by many countries from about the late 1960s. The legal basis for the grant of a Plant Breeder's Right is agreed at the international level in the Convention of 2 December 1961 (with subsequent revisions) of the International Union for the Protection of New Varieties of Plants (UPOV). Member states of UPOV are signatories to the Convention and their national plant breeders' rights legislation must conform to the requirements of the Convention as a condition of membership.

Within the EU, Council Regulation 2100/94 (known as the Basic

Regulation) establishes the concept of Plant Variety Rights as an intellectual property protection system at Community level and is operated by the EU Community Plant Variety Office at Angers, France, under the terms of Commission Regulations, notably Regulations 1239/95 (covering proceedings) and 1768/95 (on farm-saved seed). The granting of Community Plant Variety Rights (CPVR) gives protection for a plant variety throughout the EU. National systems may also be operated, but a variety cannot be protected under national and Community systems at the same time. Applicants must be based in the EU or a UPOV member country. CPVR normally confer protection for a period of 25 years (30 years in the case of vines, trees and potatoes), but for certain species and genera this can be extended for a further 5 years. During this period an annual fee must be paid by the holder of the CPVR to the EU Community Plant Variety Office.

CPVR may be granted for varieties of any species or genus but, to be eligible, varieties must be Distinct, Uniform and Stable (see above). Eligible varieties must also be *novel* at the date of application for CPVR. To be considered as novel, the variety must not have been marketed or distributed within the EU for more than 1 year before the date of application. For marketing or distribution outside the EU, this time limit extends to 4 years (6 years in the case of vines and trees). There are exceptions to this for certain kinds of disposal, for example, to cope with production and multiplication in preparation for commercial marketing, or trials of material resulting from research and development work. In making an application for CPVR, the applicant must propose a variety *denomination* (VD). This is usually (but not necessarily) a recognizable name. The VD must conform to certain requirements, both for UPOV and for the EU, and must also be approved by the EU Community Plant Variety Office. The VD requirements are concerned with avoiding duplication, confusion, offence, problems with its use, or infringement of prior rights. This VD must be used by anyone who offers the variety for sale or uses it for commercial trade. The VD must be accompanied by an official description of the variety in a format specified by UPOV, showing how it differs from other varieties of the same species.

CPVR confer certain rights, including the prohibition of propagation, production, sale, or other trade in a protected variety without the authorization of the holder of the CPVR, who thus may derive benefit by permitting these to be done, usually by means of a contract specifying terms and conditions. These normally include the payment of royalties to the holder of the CPVR. The contract may be exclusive to one person or organization, or it may be non-exclusive. However, for using seed of a major agricultural crop produced on their own holdings for planting again on their own holdings, there is a Derogation from the CPVR allowing farmers to pay 'equitable remuneration' (a royalty fee), which is a reasonable amount less than that for which they would otherwise be liable. This arrangement is known as 'farmers' privilege'. CPVR can be sold or otherwise transferred by the holder; it can be surrendered, and under certain circumstances it can be ended compulsorily by the EU Community Plant Variety Office, which also operates an appeals system.

Organization for Economic Cooperation and Development Schemes

OECD (based in Paris) was originally established in 1947 to facilitate the apportionment of USA aid to Europe under the Marshall Plan. However, it later broadened its activities to encourage economic cooperation and free trade between the countries of western Europe, and it has established certification scheme standards for seed of beet, cereals, oilseeds, herbage crops and forestry reproductive material, with the aim of improving the quality of seed in international trade. The scheme for forestry reproductive material includes material for vegetative reproduction and covers four categories of material in relation to its provenance. These standards and conditions are very similar to those of the EU and are recognized by all OECD member countries (which include most of the countries in Europe). As with ECE standards for seed potatoes, these standards are useful in controlling trade in seed between countries that recognize them, particularly between countries that are EU Member States and those that are not. The OECD schemes are also helpful in situations where seed is sent for multiplication in another country. This is often done when breeders are anxious to accelerate the multiplication of a promising new variety by growing generations out of season in a more favourable climate. Multiplication of the seed under OECD conditions then facilitates its importation to the country from which the parent seed originated. Up-to-date conditions of OECD schemes and documents may be obtained from the website www.oecd.org/agr/code/

Indexing and Diagnosis in Plant Health   10

Introduction

Diagnosis in plant health refers to the process of attributing a particular cause to symptoms observed and determining its nature and, where appropriate, the name of the causal organism. This is necessary before appropriate measures can be applied to eradicate or control a disease or infestation. The process first involves detection of the presence of a pest and then its identification; that is, assigning it to a particular taxon in an appropriate classification system. Most modern systems of classification attempt to place closely related organisms near to each other in the system and in this way to reflect the phylogeny or evolution of the group. Classically, this depended heavily on morphological appearance (phenotypic characteristics), but in recent years it has been increasingly possible to place more emphasis on relationships between genotypes as determined by nucleic acid analysis. Good classification and correct identification open access to the accumulated knowledge on the species or taxonomic group to which the pest belongs, because such knowledge is normally recorded in relation to the name of the species or other taxon. Without a name, access to this knowledge is difficult or impossible. Correct classification of the pest, both originally, when first described scientifically, and again when detected as causing a plant health problem, is therefore extremely important. However, the importance of taxonomy currently tends seldom to be recognized. The discipline is too often seen as an academic backwater and is often under-resourced, especially by hard-pressed fund-holders coping with short-term problems.

The application of modern methods of classification, particularly the use of nucleic acid analysis, to the organisms previously regarded as fungi has resulted in redefinition and a major reclassification (Waller and Cannon, 2002). The *Fungi* have been placed in a kingdom of their own, while two groups containing important plant pathogens have been separated into the kingdoms of *Protozoa* (including the genera *Polymyxa* and *Spongospora*) and *Chromista* (including the genera *Phytophthora* and *Pythium*). However, for the purposes of diagnosis these can be regarded as fungus-like organisms and are considered here together with the true fungi.

The routine testing or assessment of material for infection or infestation with pests, especially if it is to be used for propagation purposes (see Chapter 8), is often referred to as *indexing*. Even if symptoms are not present, the material may be indexed in case it harbours a latent or cryptic infection with a harmful organism.

There are a large number of different methods available for detection, testing and indexing for pests, depending on the material and the primary purpose of the exercise. Ideally, the method should be accurate, sensitive, reproducible, rapid and cheap. It may also be advantageous to be quantifiable. Any of these desirable characteristics may become more or less important according to the objective of the exercise. Pests may also be present without causing any symptoms, either latently, when symptoms may develop at a later stage, or cryptically, when the pathogen is quiescent or when its activities do not give rise to symptoms. Pests may also be present as superficial contaminants and not as parasites. The methods used should therefore be able to detect pests in whatever state they are present in or on the material examined. With so many factors to consider, the methods must be chosen carefully to be efficient, and appropriate for the material and the circumstances. For each group of pests there is a large number of different methods for detection and diagnosis, and the relevant literature is voluminous and diverse. The diagnostic techniques most commonly used and their various merits and disadvantages are summarized in Table 10.1. Scientific papers on diagnostic topics tend to be very specialized and focused on individual techniques or pests. Works that provide useful reviews of techniques for many different categories of pests include Fox (1993), Skerritt and Appels (1995) and Marshall (1996).

Detection and Isolation Methods

Methods of detection often also include a means of identification. For example, this can apply in the cases of visual examination or of inoculation to test plants for virus detection. However, often it is necessary to isolate or extract the pest from its host or substrate before

Table 10.1. Summary of common detection, diagnostic and indexing techniques for plant pests.

Methods	Useful for	Advantages and disadvantages
Visual inspection	Monitoring traded material, field checks and preliminary investigations of most types	Quick, but will not detect latent infections and often requires laboratory confirmation or further investigation
Light microscopy	Small/few samples; invertebrates, fungi, bacteria	Good for classical morphological methods of diagnosis; unsuitable for large numbers of samples, or for viruses, viroids or phytoplasmas
Electron microscopy (transmission)	Small/few samples; viruses, phytoplasmas	Rapid results for virus diagnosis; unsuitable for large or numerous samples
Electron microscopy (scanning)	Small/few samples; checking morphological details of microorganisms, especially invertebrates and fungi	Unsuitable for large or numerous samples; lengthy preparation
Serological slide tests and kits	Tests for specific pests, especially in the field and as back-up to visual observation	Quick and simple; less sensitive than ELISA; sometimes unreliable or difficult to use by inexperienced personnel
ELISA	Large/numerous samples; all antigenic pests, especially viruses, bacteria, some fungi	Good sensitivity, depending on antiserum used; rapid and accurate results; can be automated and standardized; suitable for less sophisticated laboratories
cDNA probes	Tests for specific pests, especially viroids, phytoplasmas, non-antigenic pests	Very specific, according to design of probe; lengthy and sophisticated preparation needed; unsuitable for less sophisticated laboratories
PCR	Tests for specific pests present in low concentration	Extremely sensitive and specific; can be partly automated for routine throughput of many tests; unsuitable for less sophisticated laboratories. Extraction of nucleic acid and electrophoresis are laborious
Fluorogenic 5'-nuclease assay	Numerous samples; tests for specific pests present in low concentration	Extremely sensitive and specific; can give quantitative results; can be automated; less vulnerable to contamination than basic PCR; nucleic acid extraction is laborious. At present is expensive
RFLP	Tests for specific pests and investigative tests	Can detect small genetic differences within a species; laborious, slow and unsuitable for large/numerous samples
Test plants	Confirmatory or investigative tests; virus tests in simple laboratories	Can indicate presence of unknown viruses; cheap but slow and sometimes not specific
Nutrition profiles	Identification of bacteria (classical method)	Reflects classical bacteriological taxonomy; slow to achieve conclusive results
Fatty acid and protein profiles	Identification of bacteria and certain invertebrates	Rapid results; can be semi-automated for routine throughput of many tests

cDNA, complementary DNA; ELISA, enzyme-linked immunosorbent assay; PCR, polymerase chain reaction; RFLP, restriction fragment length polymorphism.

the process of identification can start. Some of the most common methods for detection and isolation of various kinds of pests are outlined below so that their merits and limitations may be understood, but no attempt is made to give practical details for performing tests. The choice of methods used will be influenced by the symptoms observed, the type of material and knowledge of the pests likely to be present.

Invertebrate pests

Except for most nematodes, the majority of these pests are individually visible to the naked eye. They are much larger than most microorganisms. Sometimes it will be possible to identify them visually directly or with the aid of a lens or microscope, but few quarantine pests can be identified with a lens alone. If not readily visible, they may have to be extracted from the host plant or substrate by dissection or some form of extraction or trapping process. This can include sieving, either dry or wet, or flotation, as often used for the extraction of the cysts of cyst nematodes from soil. Because many countries have statutory regulations that require testing soil for these pests, this process has been highly developed, and in some countries it has been semi-automated to facilitate processing large numbers of soil samples.

Free-living nematodes are usually extracted from plant material by immersing in water or under a water spray or mist. The nematodes are collected in funnels or sink to the bottom of water columns, from which they can be removed by means of a tap. Another common method of nematode extraction for motile stages is to place the material on paper tissue supported on wire mesh. When this is placed in water so that the base of the sample is wetted, the nematodes move through the wide pores of the paper tissue and sink to the bottom of the water container. A summary of the numerous available methods is given by Barker and Davis (1996).

Extraction of invertebrates from substrates and from the atmosphere can often be facilitated by using light to attract the target pest. Many insects, particularly aphids, are attracted by yellow-coloured light or surfaces. For monitoring purposes, especially in protected environments such as glasshouses, yellow sticky traps are often used, which operate in the same way as domestic 'fly papers'. For thrips, blue sticky traps are more effective. In the field, particularly for aphid monitoring in the operation of seed potato certification schemes (Chapter 8), yellow dishes filled with water form effective aphid traps (the Moereke trap). Light traps are much used for monitoring night-flying insects, particularly pests that are moths in the adult stage. The Rothamsted light trap (Williams, 1948) is used worldwide for this purpose. Aphids, other small insects and arachnids which drift in the atmosphere without strong control of their direction of flight, can be monitored very well, both by species and by time, by means of the Rothamsted suction trap, which collects air and associated aerial fauna at the top of a 12.2 m (40 ft) pipe (Woiwod and Harrington, 1994). This system also permits the monitoring of viruliferous and pesticide-resistant insects. However, the traps normally need daily attention and considerable input of time from specialized personnel for sorting and identifying the catch.

In recent years, the use of pheromone traps has much increased, as the chemical nature of the sex attractants, or pheromones, of various pests has been determined. Synthetic pheromones (or their analogues) of target pests can then be used to attract and trap the pest selectively and efficiently from the atmosphere.

Fungi, bacteria and phytoplasmas

Species of fungi, fungus-like organisms and bacteria, which cause plant health problems, may often be found sporulating on, or oozing from, the surface of the host plant, product or substrate. If no growth is visible, or if fungal structures are immature or sterile, incubation of affected material overnight or for a longer period in damp conditions (such as on moist paper in a plastic box or Petri dish) may encourage growth on outer or cut surfaces. In such cases, a pure culture may often be obtained by transferring a few spores, mycelial fragments or cells to a culture medium with a sterile inoculating needle. Otherwise, after it has been surface sterilized, small samples of the affected material may be excised, placed in or on a suitable culture medium, and incubated to encourage growth of the pathogen from the excised sample into the medium. Mixed cultures can be purified by subculturing from selected colonies or by means of techniques permitting single-spore or hyphal-tip isolations. However, often it may be possible to examine the spores and other structures of a visible fungus or fungus-like organism directly, and make an identification without culturing. Certain plant-pathogenic organisms have so far proved impossible to culture on artificial media and are known as obligate parasites. Amongst the fungi and fungus-like organisms, these are notably the rusts, smuts, white rusts (*Albugo* spp.) and the downy and powdery mildews. In the laboratory these obligate parasites must be grown on living plants, detached leaves or leaf tissue, or on host plant cell cultures, employing special techniques.

A great diversity of both solid and liquid culture media have been developed. For solid media a setting agent is used (usually agar agar, extracted from certain seaweeds and used because microorganisms cannot normally use it as a nutrient). After formulation and sterilization, the medium is dispensed to set as a flat layer in containers with lids such as Petri dishes, bottles or jars, or to set as sloped surfaces in test tubes plugged with cotton wool or with other covers. Liquid media are usually contained in flasks or bottles, which must be shaken or stirred to ensure even growth. Cultures are usually held in incubators, which can maintain temperatures warmer or cooler than ambient, and which often also have a facility for irradiating cultures with light of various wavelengths. Many fungi are encouraged to sporulate under irradiation with 'black' light (the longer wavelengths of the ultraviolet spectrum). Certain general-purpose nutrient media, such as potato dextrose agar, malt extract agar or bacteriological nutrient agar, support the growth of a great number of different fungi and bacteria. However, special formulations have been developed to suit many particular groups, often because they have specialized nutritional requirements.

From soil, water or similar substrates, isolations may be made by mixing a small sample of the substrate (suitably diluted or concentrated if necessary) with the culture medium before setting. Because such substrates often contain a multitude of different organisms, an antibiotic or other inhibitory substance is usually added to the medium to suppress or reduce the growth of unwanted species, and thus make it easier to detect the target organism. Lowering or raising the pH of the medium may also have this effect. Many such selective media have been developed for detection and isolation of different fungi and bacteria, and are often very effective. For example, a semi-selective medium permits the direct detection and isolation of the potato brown rot bacterium, *Ralstonia solanacearum*, from river water (Elphinstone *et al.*, 1998). Summaries of methods available for detecting pathogenic fungi, fungus-like organisms and bacteria are given by Miller (1996) and by De Boer *et al.* (1996a).

Another method of selective isolation or detection is to employ a bait. This is a substance, tissue or plant that is preferentially colonized or infected by the target organism and from which it can then be cultured, identified or assessed. For example, the Gross–Gerau method for detection of *Beet necrotic yellow-vein virus* in soil employs sugarbeet seedlings as bait for the zoospores of the vector fungus, *Polymyxa betae*. If these carry this, the rhizomania virus, it will infect the beet seedlings and can then be detected with appropriate methods. A wide variety of materials are used as baits, some of the most common being pieces of apple fruit tissue and cooked or sterilized seeds of various kinds.

Phytoplasmas are distantly related to the bacteria but do not possess an outer cell wall (Jones, 2002). They are very small, restricted to the vascular tissues of the plant, and because they can alter shape due to lack of a cell wall, they are able to pass through filters with pore sizes impermeable to bacteria. Phytoplasmas cause many plant diseases, but are often difficult to detect because they are, as yet, not possible to culture on artificial media, are sparsely distributed in their hosts, and the symptoms they cause are often long delayed and not clear cut. They can be maintained as infections in certain plant species that are good general hosts, such as the periwinkles *Catharanthus roseus* or *Vinca major*, which can also be used for detection by inoculation using grafting or transmission by dodder (parasitic climbing plants in the genera *Cuscuta* and *Cassytha*). Detection by electron microscopy or UV light microscopy (using nucleic acid stains such as the fluorescent DAPI (4',6-diamidino-2-phenylindole) stain) is also possible. However, molecular methods employing a general DNA probe or PCR (see below), now make detection much easier.

Viruses and viroids

A large majority of plant viruses consist of particles with a protein coat around a core of single-stranded RNA, although a few contain double-stranded RNA or DNA (Waller, 2002). The nucleic acid contains the necessary genetic information for self-replication by subverting the genetic and cellular workings of the host cell. With some viruses the total complement of nucleic acid (the *genome*) is divided between two or more types of virus particle. They cannot be cultured outside the host cell. Often the symptoms displayed by the host will suggest whether the causal agent might be a virus. Virus detection methods for general indexing, where specific viruses are not being targeted, include inoculation to test plants (see below), transmission electron microscopy, serology using broad-spectrum antisera (see below), and molecular techniques that detect the presence of viral nucleic acid.

Viroids consist of single, circular strands of RNA without a protein coat and cause several important plant diseases, including potato spindle tuber and chrysanthemum stunt (Mumford *et al.*, 2000). They are the smallest known self-replicating plant pathogens. Pathogens of similar structure but which rely on 'helper' viruses for replication have been named 'virusoids' (Holliday, 1998). Viroids are very contagious and also relatively resistant to heat, replicating most rapidly at temperatures warm enough to inhibit replication of most viruses. The lack of a protein coat deprives them of antigenic properties, so they cannot be detected serologically and test plant inoculation or molecular methods must be used.

Testing Methods

Common techniques used by plant health diagnosticians are described below. The descriptions are necessarily very brief and cannot cover the vast range of different methods available, many of them specialized and developed for certain groups of organisms. These descriptions are not intended to be comprehensive, but to give an idea of the kind of techniques that may be employed and to explain why some techniques may be more expensive or time-consuming than others (Table 10.1).

Visual inspection

The simplest method of testing material for pests is visual inspection. Normally this will be done in most cases, even if other methods are also employed. In practice, visual inspection will only be omitted when the indexing system is automated or where large samples of items such as tubers have to be indexed and where overt symptoms are unlikely to be seen. However, although visual inspection is useful as a preliminary screen, it will very often have to be supplemented by another method because it may not be sufficiently sensitive or specific. Cryptic infections will always be missed, and it cannot distinguish between morphologically similar organisms, such as between pathological or physiological strains of a single species. However, in some circumstances, for example, where field crops have to be examined for the presence of pests that promote reliable symptoms, visual examination of part or all of a field may be more sensitive than laboratory tests on samples that are limited to a size which can be handled economically.

The reliability and accuracy of visual examination can be increased by appropriate training for the observer and by optical aids. The simplest of these, and one that should be carried by all plant health workers who might have occasion to examine material in the field, is the hand lens. Useful hand lenses normally have a magnification of about ×10, although they can be more powerful. However, the use of higher magnifications is usually limited by the amount of light available and the very short working distances.

There are various simple types of microscopes that can be used in the field or on the dockside, but it is usually best to resort to the laboratory bench when microscopic examination is necessary. This permits control over illumination and the provision of firm support for the instrument, both of which are essential for accurate observation. However, laboratory facilities for this purpose need not be elaborate and may comprise little more than a firm table, a supply of electrical power, and simple manipulating instruments and reagents. Within the European Union, Commission Directive 98/22/EC specifies the laboratory facilities and equipment required to be available at border inspection posts.

In the laboratory, two types of optical microscope are in common use. These are the stereo-microscope and the compound microscope. The two types are complementary, the stereo-microscope being used for examination and preparation of material that may then be further examined in more detail under the compound microscope. Stereomicroscopes have two separate parallel optical systems, each serving one of the two eyepieces. The resulting image is therefore stereoscopic and, because the instruments are often used for dissections, they are also referred to as dissecting microscopes. Magnification is controlled by different objective and eyepiece lenses. It can be in discrete steps or continuous (a 'zoom' range) according to the design of the instrument, and the magnification range is commonly between $\times 20$ and $\times 80$. Illumination can be by incident or transmitted light. Although transmitted light is important for nematode examination, for much examination of plant material incident light is the more useful, especially when provided by a source that can be finely adjusted for intensity and for angles of incidence, such as the modern designs of multi-headed fibreoptic lights. A great advantage of the stereomicroscope is that specimens can be examined with a minimum of preparation, direct from the field. No sectioning or staining is normally necessary.

The magnification of stereomicroscopes is usually insufficient for detailed examination of microorganisms and for this the compound microscope must be used. Although usually provided with binocular eyepieces for professional use, compound microscopes have a single optical system. However, they can be extremely complicated pieces of equipment and a great many different facilities can be built in or fitted as attachments. For plant health work the most usual of these are probably the camera attachment, allowing photomicrographs to be made easily in colour or monochrome; the UV light system, for use with fluorescent stains or immunofluorescent conjugates; and the phase-contrast or differential interference contrast light system, for use in examining colourless, translucent structures or organisms without the need to stain them. The magnification of the compound microscope is also controlled by varying the eyepiece and objective lenses: zoom lenses are not used. The range is normally from about $\times 100$ ('low power') to about $\times 400$ ('high power'). Magnifications above this, up to about $\times 1000$, are possible with special techniques, such as oil immersion (where oil of a matching refractive index joins the objective lens to the slide cover slip), the ultimate theoretical magnification limit being controlled by the wavelength of the light used. However, in routine practice it is seldom necessary to go above the normal 'high power' magnification, and the electron microscope has superseded high optical magnifications for many purposes. For examination, specimens are normally prepared immersed in a liquid on a glass slide, with a glass cover slip. The liquid may be water, but often is a non-drying reagent with a suitable refractive index for the material to be examined, and sometimes also containing a stain or other reagent to render the specimen more easily visible. Permanent or semi-permanent slide preparations can be made. Where structural relationships between organisms or within cells are being examined, material may be sectioned, either by hand with a razor or with a microtome, a machine with which sections can be cut more accurately and much thinner. For delicate material, a careful and lengthy process of embedding in paraffin wax may be necessary before the specimen is mounted on the microtome, but a freezing microtome, which freezes the material before cutting, often permits this to be avoided with more robust subjects.

Binoculars are an optical aid not normally associated with plant health work, but which are sometimes useful. They can be helpful in field or store inspections where access is difficult, and particularly for aerial surveys. From heights above 500 m it is often difficult to be certain of the identity of a field crop and a good pair of binoculars with a wide angle of vision but moderate magnification ($\times 8$ is sufficient) can be of great help.

Electron microscopy

The electron microscope developed rapidly in the last half of the 20th century, and during that time it has been responsible for much of the great advance in knowledge of viruses and the fine structure of living cells. In essence, it works in a similar way to the light microscope but uses a beam of electrons instead of light. As with light microscopes, there are two types: the scanning electron microscope (SEM), which is used to examine the surface of objects, and the transmission electron microscope (TEM), in which the electron beam passes through the specimen to be examined. Magnifications in the region of $\times 4000$ to $\times 40,000$ are usual for the SEM and from $\times 20,000$ to 200,000 for the TEM, but can be higher or lower than this.

With the TEM the electron beam is handled and focused by electromagnetic lenses in much the same way as optical lenses focus light. Control of magnification is by adjustment of the electromagnetic lens system and observation is by visualizing the image either directly on a visual display screen or indirectly by capturing the image on photographic film. Specimens are placed on a support film on grids of fine copper-wire mesh and examination is made in the areas between the wires. There are many different methods of specimen preparation, but biological material normally must be dehydrated. For pathogens within plant cells, the cell contents can be expressed and the resulting sap sprayed on to the grids and dried before examination. To examine the fine structure of cells and their contents, sections may be cut on the ultra-microtome, after suitable treatment and embedding in resin or plastic. This produces extremely thin slices of the specimen for examination. The mounted section may be further treated to enhance observation, such as with various kinds of electron-dense stains,

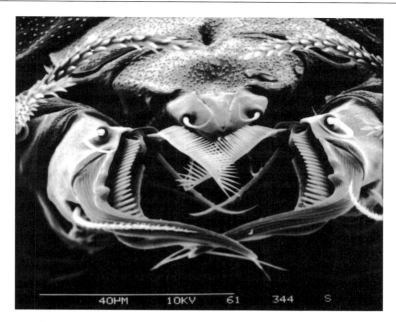

Fig. 10.1. Scanning electron micrograph showing the mouthparts of the prostigmatid predatory mite *Cheyletomorpha lepidoptorum*. Note the depth of focus. (Photo: courtesy of J. Ostoja-Starzewski, CSL.)

shadowing with gold or other material, in the same way as mentioned for the SEM below, or by labelling in some way the target to be examined. Labelling is often by means of a serological technique and will be described under that heading.

In the SEM the electron beam is scanned over the specimen, causing the emission of secondary electrons. These are used, through electronic capturing, processing and focusing, to form the image on a screen or on film. Biological specimens to be examined must be coated with a thin film of an electrical conductor such as gold or carbon. Usually this involves dehydration of the specimen and its subsequent coating by electric arc in near vacuum conditions using a sputter coater. However, in recent years methods have been developed to permit examination of material without dehydration. One of the attributes of SEM images is the great depth of focus obtained, which permits sharp images of relatively stout objects (Fig. 10.1).

Serological techniques

These techniques exploit the affinity between antigens and antibodies in the immune system of warm-blooded creatures, and summaries of their use in detection and diagnosis of plant pests and pesticides are given by Barker (1996) and Dewey (2002). Antibodies are produced when an antigen, for example, a purified preparation of a plant virus or plant pathogenic bacterium, is injected into the bloodstream of a mammal or bird. Microorganisms or their organs (such as spores) vary a great deal in their ability to incite the production of antibodies, a characteristic that is associated with compounds on their surfaces, which are able to act as antigens, such as proteins or lipopolysaccharides. Some pathogens, such as viroids, which do not bear these substances, will not behave as antigens. However, others, particularly many viruses and bacteria but also a great range of other organisms and their constituents, carry good antigens and will promote the production of high concentrations of antibodies (expressed as high-titre antisera).

Laboratory animals used for antiserum production are usually rabbits, rats or mice, but other domestic species, including chickens, are sometimes used for special purposes. After an incubation period, the animal is bled and the antiserum separated from the blood. Antisera produced in this way will contain a mixture of antibodies produced by the animal's immune system in response to each of the various epitopes (molecular sites) on the antigen introduced. There may be very many different antibodies present in such a polyclonal antiserum, including antibodies produced in response to sites on impurities (such as plant cell debris) introduced at the same time as the intended antigen. Polyclonal antisera are therefore often not very specific, although their specificity can be increased (sometimes very substantially) in various ways. Besides reacting with the original type of antigen, they will also react with any other antigen, including impurities, which carry the same epitopes as those of the original antigen. For example, this could be a related virus or bacterium, or plant cell debris. This type of antiserum may therefore produce false positive results when used for diagnostic purposes. Nevertheless, provided reaction to impurities is not serious, they can be very useful, and the broad specificity of a single polyclonal antiserum can also be exploited to detect the various forms of an organism that exists in several different strains.

To overcome the problems caused by lack of specificity in polyclonal antisera, monoclonal antisera have been developed. This involves the immunization of an animal (usually a mouse or rat), removal of the spleen and culturing the spleen cells, which are a good source of the lymphocytes responsible for antibody production. These cells are then conjugated with myeloma cells, which permit unlimited propagation in tissue culture. There then follows a lengthy process of isolation and testing, perhaps of many hundreds of antibody-producing cell lines, to select those that produce antibodies with the desired degree of specificity. Cell lines producing antibodies which are more-or-less specific can be selected according to the distribution on the intended target or targets of the epitopes to which they react. The resulting monoclonal antiserum can be very specific, even to sub-specific strains, and eliminates cross-reactions with other antigens. Specificity can be widened, if desired, by careful selection for reaction to epitopes carried by all the strains or other variants it is desired to detect, or by combining several lines into one composite antiserum. Monoclonal antisera thus not only have good specificity, but also have the advantage that, once developed, the cell lines producing the antibodies can be cultured *in vitro* indefinitely without further use of laboratory animals or laborious purification procedures.

Once produced, antisera can be used in a great variety of different serological tests. These often depend on the ability of antibodies to be attached to inert substances, such as the surface of plastic wells or latex spheres, and to various dyes and enzymes. It is important to trial serological methods thoroughly before adopting them for routine use, as the various methods, reagents and individual laboratory animals can vary considerably and some antisera will be found unsuitable for use with certain methods or for detecting certain antigens.

The simplest serological tests often depend on the agglutination or clumping together of antibody-labelled cells or beads in the presence of the target antigen to give a visible pattern. A major advance in the use of serological techniques for detection and identification of plant pathogens came with the development of the ELISA by Voller *et al.* (1976) and its use for detection and identification of plant viruses by Clark and Adams (1977). Subsequently this has been refined, developed and automated, and it remains one of the main techniques for detection of antigenic pests in plant diagnostic laboratories.

A great many variants of the ELISA have been developed for routine use or for special purposes. The ELISA is normally performed in small wells, of which the usual kind of microtitre plastic plate contains 96. One of the most common types of ELISA is the double-antibody sandwich or DAS-ELISA, illustrated by Fig. 10.2. For this the interior surface of the wells is first coated with antibodies to the target organism or other antigen by introducing a suitable antiserum and washing away the surplus. A polyclonal antiserum is usually used for this purpose as specificity is not critical and polyclonals tend to trap the target more effectively. The sample to be tested is then introduced, usually in the form of plant sap or an aqueous suspension, suitably diluted with a buffer solution to control the pH. Any target antigens present in the sample will attach to the antibodies on the walls of the well. After a suitable incubation period, the sample is washed away and a second antiserum to the target antigen is introduced. This second antiserum may be a monoclonal and is linked to a suitable enzyme (usually alkaline phosphatase). It will adhere to any target antigen trapped by the first antibody on the well walls. Finally, after a further washing to eliminate

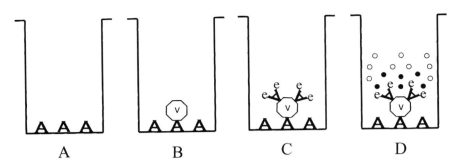

Fig. 10.2. Diagrammatic representation of the double-antibody sandwich-ELISA. (A), specific antibody, **A**, adsorbed on to well walls; (B) target virus, v, from test sample attached to antibody; (C) antibody labelled with enzyme, e, attached to virus; (D) enzyme substrate, ○, by action of enzyme gives colour, ●. (Redrawn from Clark and Adams, 1977.)

any antibody–enzyme conjugate not attached to the target antigen, a substrate for the enzyme (*p*-nitrophenyl phosphate for alkaline phosphatase) is introduced and the plate is incubated under standard conditions. If any target antigen is present, the action of the enzyme on the substrate produces a coloured compound (yellow with alkaline phosphatase), which can be detected visually or with a colorimeter. Both positive and negative controls are included on each microtitre plate. The whole process can be completed within 24 h or less, and the system can be automated to deal with large numbers of samples. A further advantage is that the concentration of colour can be measured photometrically to give an indication of comparative concentrations of the target antigen present.

In one common variant of the DAS-ELISA, the second antibody applied is one produced in a different animal species to that of the antibody on the well walls and is not conjugated with the enzyme. Instead, the enzyme is conjugated with an antibody produced in response to the second antibody (an anti-antibody antibody) in a different species of animal. This will attach to any antibody of the particular animal species used and so will attach to any such antibody attached to the target antigen but not to the surplus antibody coating the well walls (Fig. 10.3). Subsequent incubation with the enzyme substrate will then give the same colour reaction as before. This type of ELISA is known as the indirect DAS-ELISA and has the advantage that the same antibody–enzyme conjugate can be used for detecting different target antigens, provided the second applied antibody has been produced in the requisite species of animal. Another advantage is that enzyme-labelled anti-mouse and anti-rat secondary antibodies are available commercially, thus saving busy diagnosticians both time and cost.

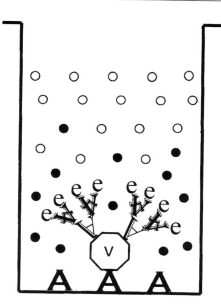

Fig. 10.3. Diagrammatic representation of the final stage of indirect double-antibody sandwich-ELISA. The target virus, v, is trapped to the well walls by virus-specific antibody **A**. The virus-specific antibody, A, which has been raised in a different animal, has also attached to the virus. The enzyme, e, is linked to antibody A (which is specific for antibodies raised in the same animal species as A) which has linked to A. The enzyme substrate (○) gives colour; (●) by action of the enzyme.

In general, serological techniques are much more widely and successfully used to detect and identify viruses and bacteria than other kinds of microorganisms. This is due to many factors, but mainly because organisms much larger and more complicated than viruses or bacteria carry many more antigenic substances, and also the antigenic properties of their different life stages may vary. Also, whereas viruses have been classified partly on their serological relationships, most other organisms have not. However, successful serological tests for many fungi and fungus-like organisms, and for certain life stages or characteristics of some insects, nematodes and other invertebrates, have been developed.

Serological techniques are sometimes used to enhance the performance of electron microscopy. In such immuno-electron microscopy, antisera can be used on the electron microscope grid to trap antigens such as virus particles from a suspension that might otherwise be too dilute for their detection. Where preparations for electron microscopy contain a mixture of different virus particles of similar morphology, or where it is desired to identify specifically a virus particle under the electron microscope, the electron microscope grid bearing the sample can be treated further with an antiserum to the target organism. When examined in the electron microscope, or in the resulting electron micrograph, the target organism will then appear to be 'decorated' with a fuzzy coat of antiserum, thus specifically identifying it, whereas other particles will remain uncoated. In a variant of this method, the decorating antibodies can be attached to colloidal particles of gold, which then appear as solid black specks in the fuzzy coat, making identification even more positive (Fig. 10.4).

For bacteria, the immunofluorescence detection method is one of the most sensitive to have been developed. This procedure depends on staining bacterial cells supported on a microscope slide with specific antibodies conjugated with a fluorescent dye. The antibodies will

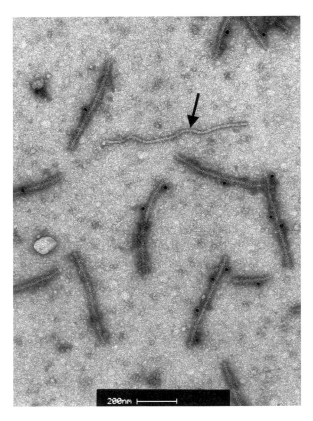

Fig. 10.4. Immunosorbent transmission electron micrograph showing virus particles of *Potato virus X* (PVX) and *Potato virus Y* (PVY). The preparation has been treated with antibody to PVX linked to protein-A gold particles. The PVX particles appear decorated with antibody and associated with the gold particles (the black dots). The PVY particle remains undecorated (arrow). (Photo: courtesy of Daphne Wright, CSL.)

naturally attach to the type of bacteria originally used to raise the antibodies and not to others. These are then observed at high magnification (usually ×1000) under a compound microscope either designed specifically for fluorescence work or fitted with attachments for this. High-intensity light of the wavelengths necessary to incite fluorescence of the dye is transmitted to the subject via an optical pathway through the objective lens, causing the bacterial cells that carry the fluorescent antibody conjugate to glow with fluorescent light, rendering them easily distinguishable. The method has also been used successfully to detect and identify other biological structures, such as fungal spores, and can be made quantitative by counting at different dilutions.

Serological methods have also been developed to detect and identify pesticides and mycotoxins, especially in foodstuffs. Pesticide and mycotoxin molecules do not normally behave as antigens on their own, but they can be linked to an antigenic carrier, such as a protein, to become *haptens*. These can be used with standard techniques to immunize animals and produce antisera for use in immunoassays. However, although in many instances the sensitivity and specificity of such tests are equal to, or better than, conventional methods, there are disadvantages, which have limited their general adoption. For example, antibodies to pesticide haptens often cross-react with chemical analogues and may be more sensitive to some pesticides than others. This is obviously a disadvantage when it is desired to analyse foodstuffs simultaneously for several different pesticide residues, as is often the case. Also, serological tests such as ELISA are unsuited to the simultaneous detection of many different antigens.

A considerable number of different kinds of immunoassay kits have been developed commercially for use by people without specialized knowledge, in various different situations, to detect different microorganisms. Most of these are for detection of virus pathogens, but kits for fungal and bacterial pathogens are also available. They are intended to be used for targeted monitoring and quick, on-the-spot confirmation of visual assessments, but do not obviate the need for visual inspection. Although many of these kits work well, they are limited to the specific purposes and pests for which they were designed. They have to be used with caution, and for plant health work they are used mostly in support of field inspections for disease incidence relating to decisions on pesticide application or assessments for export, import or certification of healthy stock. However, it is important to be aware that even the most sensitive methods are limited by the degree to which samples are representative of the material or crop being tested.

Most kinds of immunoassay kits are not suitable for investigation of samples in the laboratory or for definitive diagnosis. Lateral flow devices are an exception to this. These devices were originally developed for

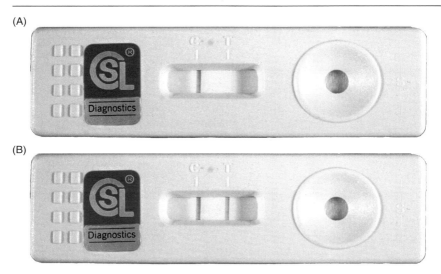

Fig. 10.5. Lateral flow device for serological diagnosis of viruses and other pests: (A) negative; (B) positive. (Courtesy of CSL. Crown copyright.)

pregnancy testing in the home and the principle has been developed for use in testing for plant pathogenic viruses, bacteria and fungi. They are based on the flow of plant extract, buffer, and antibody-coated coloured latex beads across a membrane impregnated with two lines of trapping antibodies. Results are displayed within 2 min in a window as visible coloured lines (Fig. 10.5). One line of antibodies is specific for the target pathogenic organism; the other is specific for a constituent of the plant species being tested. Two visible lines therefore indicate a positive result, while a single line (detecting only the plant species and acting as a check on the correct working of the device) indicates no infection. The sensitivity and accuracy of this method is comparable to visually assessed ELISA tests. Although primarily intended for use in the field, they also provide a quick and accurate laboratory test that is cheaper than ELISA when only a small number of samples is to be tested.

Molecular techniques

Molecular techniques in plant health diagnostic work depend on the identification of characters in the actual genetic make-up (the genome) of the target organism. The genome is composed of nucleic acids, which may be RNA (in the case of most plant viruses) or DNA. All organisms with living cells contain DNA and this is composed of a sequence of nucleotides. As with serological techniques, very many different molecular diagnostic methods and variants of them have been developed and this field continues to expand rapidly (Mills, 1996; Bridge, 2002).

Nucleic acid probes

One of the most useful methods is the use of nucleic acid *probes*, consisting of a known sequence of nucleotides, constructed or selected for their ability to link to (*hybridize* with) a unique complementary sequence of nucleotides present within the target organism's nucleic acid. The probe may be regarded as a 'key' that identifies the target organism by fitting into that organism's nucleic acid 'lock'. The complementary nucleic acid probe (which can be either DNA or RNA) is labelled in some way to make it detectable after hybridization, often with a radioactive phosphorus isotope (^{32}P), which can be detected by autoradiography, or with an enzyme or other chemical that produces a coloured substance when treated with a suitable reagent. Although radioactive labels are very effective, they have serious disadvantages in that the shelf-life is short (due to the short half-life of ^{32}P), they present a safety hazard and they require costly disposal procedures. More recent developments have therefore favoured non-radioactive labels, which now can be equally sensitive and give more rapid results.

The hybridization reaction is usually performed with the target nucleic acid bound on to nitrocellulose paper, nylon, or other plastic surface. This format is particularly convenient when many samples are to be tested (Mumford *et al.*, 2000). To do the test, nucleic acid may first be extracted from the sample or, if suitable, the sample can be applied directly, for example, as diluted plant sap. Various techniques have been developed to extract nucleic acid from difficult substrates, including soil, often by employing some kind of capture method. For example, the immunocapture technique uses an antibody-coated surface to trap and concentrate antigenic entities, such as viruses, before molecular methods are performed.

The polymerase chain reaction

The sensitivity of nucleic acid probe methods of detection is comparable to that of serological methods. However, sensitivity can be greatly increased by multiplying either the target DNA sequence or the probe. This can be done by means of the PCR, which relies on the action of a DNA polymerase enzyme, usually one named *Taq* isolated from the thermophilic bacterium *Thermus aquaticus*, which is found in hot springs. It is also necessary to have oligonucleotide primers, which are lengths of nucleic acid containing sequences of about 20 nucleotides complementary to those on either side of the target DNA sequence to be multiplied. In effect these will demarcate the target DNA sequence. The reaction mixture must contain appropriate proportions of double-stranded DNA with target sequences from the organism to be detected, the oligonucleotide primers, deoxynucleotide triphosphates (for assembling new lengths of DNA), the polymerase enzyme and a suitable buffer.

First the mixture is heated to 94°C. This denatures the double-stranded DNA, separating the two strands. The mixture is then cooled to a suitable temperature (between 30 and 72°C), which will allow the oligonucleotide primers to link (anneal) to the complementary sequences on each of the separated DNA strands. The mixture is then heated to 72°C, at which temperature the polymerase promotes the assembly of new DNA strands with nucleotides complementary to those on the target sequence between the primers, thus resulting in two new lengths of double-stranded DNA similar to that of the original (Fig. 10.6). The thermal cycle is then repeated. This is usually done between 25 and 40 times, using a thermal cycler machine. The quantity of DNA is doubled after each cycle until one of the reaction components is

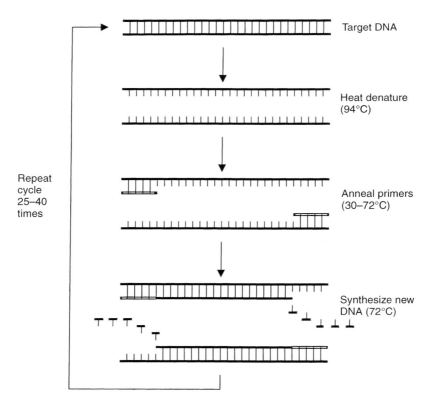

Fig. 10.6. Diagrammatic representation of the PCR process (redrawn from Mills, 1996).

exhausted or the polymerase is denatured and becomes inactive. The products of the PCR process (mainly the multiplied quantity of the target sequence of DNA) must then be detected and analysed. This is normally done by electrophoresis in agarose gels, which separates DNA fragments into groups of similar length that form bands in the gel. These bands can then be made visible by treatment with a suitable dye and identified by comparison with a standard control run alongside (Fig. 10.7).

For organisms with RNA genomes (the majority of plant viruses and viroids) the target RNA nucleotide sequence must first be transcribed into a complementary sequence of single-stranded DNA before PCR is done. This reverse transcription employs an enzyme such as AMV reverse transcriptase and is normally done at 42°C for 30 min before the standard thermocycling is begun. The process is known as reverse transcription PCR, or RT-PCR.

There are many ways in which molecular methods of detection and diagnosis can be enhanced. For example, the sensitivity of the PCR can be increased using a technique known as 'nested primers'. In this technique the PCR exercise is repeated using a second pair of oligonucleotide primers, which attach to sites within the sequence demarcated by the first pair. This second PCR exercise thus multiplies only a part of the sequence length of DNA that was multiplied by the first PCR, but the final degree of multiplication will be much greater. Use

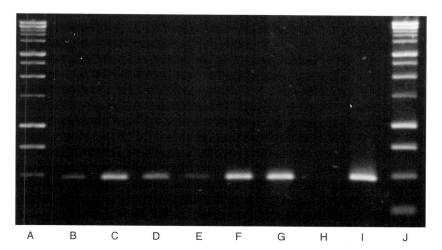

Fig. 10.7. An agarose gel after electrophoresis, showing the detection of *Tobacco rattle virus* using reverse-transcription-PCR. The primers used were designed to amplify a 463 base pair product from within the RNA-1 part of the virus genome. Gel lanes B to G show six positive samples, while lanes H and I show negative and positive controls, respectively. Lanes A and J contain standard DNA size markers. (Courtesy of Dr R. Mumford, CSL.)

of this technique for the detection and identification of *Phytophthora* species is described by Duncan and Cooke (2002).

Although PCR is an extremely useful technique, its very sensitivity can be a disadvantage in that the slightest contamination of samples can give false positive results. It therefore needs an ultra-clean environment to perform well. This, and the fact that it is easily inhibited from working properly by impurities which may be extracted together with the target DNA, render it unsuitable for use in less sophisticated laboratories where serological methods, especially ELISA, are usually more reliable. The need to perform electrophoresis for identification of the PCR products also means that it cannot be fully automated and is costly in labour. It is therefore unsuitable for applications requiring a regular throughput of large numbers of samples.

The fluorogenic 5'-nuclease assay

In the mid 1990s techniques were developed that give immediate detection of PCR products. The most successful of these is the fluorogenic 5'-nuclease assay, which has been patented commercially, using probes marketed under the trademark TaqMan®. This uses the fact that the PCR *Taq* polymerase also has 5'-nuclease activity. The mechanism is illustrated in Fig. 10.8. In the fluorogenic 5'-nuclease assay, primers are used that demarcate a relatively short length of nucleic acid. In addition, a nucleic acid probe is used that anneals to a sequence in the target nucleic acid between the primers. The probe is labelled at the 5' end with a reporter (fluorescent) dye and at the 3' end with a quencher dye. While the probe is intact, the proximity of the quencher eliminates the detectable fluorescence emitted by the reporter dye but, when the probe is cleaved by the 5'-nuclease activity of the *Taq* enzyme, the two dyes are separated and the fluorescence becomes easily detectable (Fig. 10.8). With each PCR cycle, more of the probe anneals to the target nucleic acid, which has been produced and during the next cycle this is in turn degraded by the *Taq* enzyme, eventually increasing fluorescence to detectable levels and beyond. The greater the amount of target nucleic acid present initially, the earlier the cycle at which fluorescence will reach detectable levels. Thus the fluorogenic 5'-nuclease assay can not only immediately signal the presence of target nucleic acid by means of fluorescence, but can also deliver a measurement of its quantity. Because the need to prepare and run electrophoretic gels is eliminated, the system can be fully automated and is much quicker and cheaper than the basic PCR technique. It can be adapted to a variety of applications (Mumford *et al.*, 1999). It is also less prone to inhibitors and is extremely sensitive, being about 100 times more sensitive than the standard PCR test and up to 10,000 times more sensitive than ELISA. The fluorogenic 5'-nuclease assay is performed in

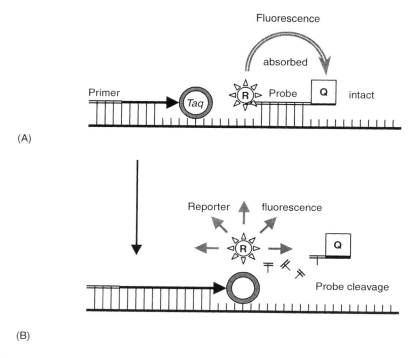

Fig. 10.8. Diagrammatic representation of the fluorogenic 5′-nuclease assay process. (A) Polymerization starts from the primer and progresses towards the TaqMan® probe, which is intact. (B) Polymerization reaches the probe, which is cleaved by the *Taq* enzyme, allowing the fluorescence to become detectable. (After Rebecca Weekes, CSL, unpublished.)

wells in microtitre plastic plates similar to those used for ELISA. Once filled with the sample and reagents, the wells can be sealed with a cap, which greatly reduces contamination. Nevertheless, in common with other PCR techniques, the fluorogenic 5′-nuclease assay requires extraction of nucleic acid from the sample under test, which at present is labour intensive and therefore costly.

Restriction fragment length polymorphism

A technique that can be used either directly with extracted DNA or after multiplication of target sequences by PCR is restriction fragment length polymorphism (RFLP) analysis. The DNA is digested with one or more endonuclease enzymes, each of which cleaves the DNA at specific sites. This cuts the DNA into a range of fragment sizes, which can be separated by electrophoresis in agarose gels to give a characteristic pattern of bands. This pattern of grouped fragments of similar size can be detected by staining with ethidium bromide or by hybridization with a suitably labelled DNA probe. In organisms more complicated than viruses or viroids, nucleic acid is found in several different types of organelle within the cell, in addition to the chromosomes of the cell nucleus. These include the ribosomes, the mitochondria and, in bacteria, the plasmids. The nucleic acid in each of these structures has different characteristics. For example, RNA has several conserved regions that are found universally in all living cells, but these are interspersed with shorter regions, which vary from organism to organism. There is much more nucleic acid (DNA) in mitochondria than in ribosomes and the variable regions are much larger. RFLP analysis of nucleic acid from ribosomes will therefore give different results to those from mitochondria and may be more or less suitable for the identification of particular organisms. By using universal primers that attach to the conserved regions, the variable regions of ribosomal nucleic acid can be multiplied without knowing what kind of organism may be present. Subsequent RFLP analysis can then identify the organism. However, this method is often less exact than RFLP analysis of mitochondrial DNA, which results in a much greater variety of nucleic acid fragments. If no specific fragment group unique to the target organism can be used for diagnosis, one or more probes linking to a few or many fragment groups characteristic of the organism can be used to give a diagnostic pattern.

Bioassays

Inoculation to test plants and subsequent observation of the distinctive symptoms developed was one of the first methods to be used for the identification and characterization of plant viruses. By trial and error, a number of plant species have been identified as particularly suitable for this purpose. This is because they are: (i) susceptible to infection by a very wide range of different viruses; and (ii) they develop symptoms that are often distinctive for particular viruses, either alone or when taken together with those on other test plants. On herbaceous test plants these symptoms may be either systemic or localized on certain organs, usually the leaves that have received the inoculation, where they often appear as discrete, local lesions. In this case they can also be used as a quantitative assay. Such test plants very commonly include *Chenopodium amaranticolor*, *C. quinoa*, *Cucumis sativus* (cucumber), *Gomphrena globosa*, *Nicotiana clevelandii*, *N. debneyi*, *N. glutinosa*, *N. tabacum* (tobacco), *Phaseolus vulgaris* (common bean), *Physalis peruviana* (Cape gooseberry) and *Vicia faba* (broad bean). Certain varieties or strains of test plant species are often better indicators than others and should be specified when referring to such tests.

Inoculation is usually by gently rubbing the leaves with an aqueous suspension of infective propagules mixed (in the case of viruses) with an abrasive such as carborundum powder or celite. With bacteria it is often by injection or introduction to wounds, and with fungi and funguslike organisms often by external application of spores or a mycelial culture. With viruses and phytoplasmas, which are systemic in the host, and where other methods are ineffective, inoculation may be by various forms of grafting or by means of the dodder, which can be induced to parasitize both donor and test plants and so effect transmission. Woody plants, especially fruit trees, are often tested by grafting or budding to woody indicator plants which are themselves fruit or ornamental trees of certain varieties particularly sensitive to infection. In these cases the delay between inoculation and reading of any symptoms developing may be as much as a year or more, whereas with herbaceous test plants the interval is usually a matter of about 2–5 weeks. The longest delay is where symptoms can only be discerned at fruiting, and in these cases the delay may be as long as 4 years, if more than one fruit crop is required to confirm observations.

Because of the length of time needed and the improvement of other methods of testing, tests with woody indicators are now used mainly for confirmation purposes, or where it is desired to detect any as yet unknown pathogen that may be present (for example, in the production of nuclear stock). However, inoculation to herbaceous test plants is still an important and commonly used method, especially to support other methods of testing for viruses and bacteria, and as a first screen for material of unknown provenance. Plant diagnostic laboratories need a constant supply of a selection of test plant species at a suitable age for inoculation.

Physiological and biochemical methods

For bacteria the classical means of identification is by nutritional profiling and biochemical tests. For this the bacterium must first be isolated in pure culture and a suitable range of artificial media are then inoculated. By observation of the ability of the bacterium to utilize the different media, as food sources, for fermentation or the production of metabolites, a physiological and biochemical profile of the bacterium can be obtained, which can then be compared with known profiles and characteristics to identify the species. This is a relatively slow process that may occupy up to several weeks.

Other kinds of profiles can also be used for identification of bacteria and other organisms (Stead, 1995). These include fatty acid, protein and enzyme profiles, which usually separate the extracted fatty acids or proteins by electrophoresis in gels or columns. Each profiling system is suited to the identification of a different spectrum of organisms. Fatty acid profiling is particularly suited to bacteriological identification and can be automated, allowing the throughput of many samples with relatively little labour. A pure culture must first be obtained. It is suited to a great many groups of bacteria, but there are a few for which it is not useful. Protein profiling can be used for some of these. Enzymes are proteins that catalyse a vast number of biochemical reactions within biological cells and enzyme profiling is a specialized sector of protein profiling. Enzyme profiling can be useful for the identification of organisms belonging to several diverse groups, including insect larvae, such as the *Liriomyza* leafminers (Collins, 1996), certain genera of fungi and fungus-like organisms (such as *Colletotrichum* and *Phytophthora*), and for plant varieties, for example, varieties of apple and strawberry. Correct variety identification is very important in certification schemes for planting material (Chapter 8) and in cases of serious doubt this method can sometimes be used to supplement or confirm the results of visual inspection.

Uses of Indexing and Diagnosis

Support of plant health regulations

Plant health regulations governing both domestic and international trade apply to certain species or groups of plant pests, certain plant species or groups, and to some plant-related commodities and products. In practice, they may need to be applied to domestic or international trade, in emergency situations, during longer-term phytosanitary campaigns, and as part of the operation of certification schemes for planting material or the production of nuclear stock. It is therefore vital that plants and pests are identified correctly before such regulations are applied. Incorrect identification could lead to unnecessary major losses for the producer or trader, and could involve plant health authorities in unnecessary and costly lawsuits and compensation payments. Initially, the application of regulatory measures frequently depends upon the vigilance and expertise of phytosanitary inspectors and visual inspection, which often needs to be supported and supplemented by appropriate and more sophisticated diagnostic methods. The suitability of available methods for the task in hand, including speed, cost and reliability (Table 10.1), must be assessed carefully and methods chosen to suit the circumstances.

The time available for inspection and diagnosis may be very limited, especially if the material is perishable. In these situations, the diagnostic methods used must therefore give rapid results and also must not be more costly than the trade can support or justify. Quick confirmatory tests using light or electron microscopy, or a serological test, will often be appropriate. Sampling, packaging, labelling and transport to the laboratory all take significant time and, even with rapid methods of testing and perhaps the bulking of several samples together, the processing of a large number of samples can extend to several days. Where testing and diagnosis are likely to require more lengthy procedures, allowance for this should be made, where possible, perhaps by adjusting the time or place of inspections or sampling. In surveys following the discovery of an introduced pest the sampling and detection methods used must be sensitive enough to detect the pest while it is still rare, and rapid enough to give results before widespread dispersal occurs. For routine surveys, plant health authorities will usually have greater freedom to decide on their timing, extent of sampling and the diagnostic methods to be used.

Because of the importance of correct diagnosis in the application of plant health regulations and measures, international efforts have been made to agree on standard methods and protocols for pest diagnosis. The ISTA has been particularly active in pursuing this aim with regard to the testing of seeds (Chapter 8). The RPPOs, notably NAPPO and EPPO (see Appendix I), have initiated panels and programmes for harmonizing diagnostic methods and protocols for detection and diagnosis of pests affecting various kinds of plant material. In particular, an ongoing EPPO cooperative programme aims to develop and publish agreed indexing protocols for the harmful organisms identified as quarantine pests for Europe.

Support of healthy stock certification schemes

The nuclear stock is the basis for schemes producing certified pathogen-tested planting material (Chapter 8). It is therefore important that the nuclear stock is tested with the most accurate and sensitive methods available, to detect all the pests that are known to be transmissible with planting material of the host species concerned. The only exception to this is that, unless the material is imported from another region, it is not necessary to test for pests of the host that do not occur in the region where it is grown. The testing of material destined to become nuclear stock is therefore often lengthy and may use a wide variety of methods, each appropriate for detection of different pests. To guard against the possibility that viruses may be present in low concentration, or the presence of other factors that might give false negative results, tests may need to be repeated over a lengthy period (perhaps several years) in order that negative results can be viewed with confidence. Tests may also need to be done at certain specific times of the growing season and, to guard against the possibility that an unknown or unanticipated virus pathogen is present, it is also advisable to test the material on a range of suitable test plants as well as using laboratory tests.

Many certification schemes include a requirement for laboratory tests at certain stages, especially at the higher grades. With seed potatoes, for example, virus tests on samples of leaves are often required at the highest grades, while, at lower grades, post-harvest virus tests may be required on tuber samples (De Boer *et al.*, 1996b). This is particularly important where seed potatoes are grown in areas that are favourable to the rapid increase of virus vector aphid populations during the growing season. However, because cost limits the size of tuber samples that can be tested, post-harvest tests on seed tubers mainly identify those stocks that, for some reason, are of very poor virus health.

Documentation and Record Keeping in Diagnostic Laboratories

Strict discipline in the keeping of diagnostic records is essential in plant health, both to facilitate quick and accurate diagnosis and to build confidence, both nationally and internationally, in the laboratory and the expertise of its workers. It will often be necessary to satisfy queries from plant health authorities, plant growers and traders as to the progress of particular samples submitted for diagnosis. It must be possible to do this quickly and without too much expenditure of expensive scientific or administrative time. All laboratories therefore need some system for recording the receipt of samples, their progress through the testing procedures, and the outcome of the diagnosis. For this a computerized system is desirable, so long as it is simple to use and reliable. Otherwise, particularly for smaller laboratories, manual systems employing index cards and accession ledgers may be more satisfactory. Most systems operate by allotting a unique serial number to each sample received. This permits it to be traced throughout its period of diagnosis, and the history of its examination is recorded with reference to this number, either on the computer or on the paper records.

In some laboratories, particularly those dealing with the detection and analysis of pesticides, it may be appropriate to acquire accreditation by one or other of the available quality assurance schemes, such as Good Laboratory Practice (GLP) or EN45001. This is especially necessary if results are to be used in regulatory enforcement or other legal proceedings. However, it may not be essential for laboratories concerned only with diagnosis of pests on plant material. Where quality assurance accreditation is to be attained, it must be accepted that this will considerably increase costs. Even if this is not attempted, diagnostic procedures should adhere to written SOPs, which accurately describe the protocols, reagents and equipment used. This will promote consistency in results and will permit cross-checking where results vary.

The use of SOPs produced by international organizations provides a basis for international comparisons.

Some kind of archiving for records and material will be necessary, especially for quality assurance schemes. For records, this can be done electronically or as hard copy. For specimens, this may be needed for both the short term (for example, to guard against the need for testing or re-testing more material of the specimen received) and the longer term, for reference purposes or for evidence to respond to challenges on diagnostic results. There are a great number of techniques for preserving material, living or dead, according to its nature. For example, diseased plant material and fungus cultures may be dried and mounted on paper sheets or in folders, or preserved in alcohol or other liquid preservative. Fungal or bacterial cells can be freeze dried and preserved in glass capsules for later resuscitation, and viruses and bacteria can be preserved in a viable state by deep freezing in freezers or in liquid nitrogen. Fungus cultures can also be kept viable in long-term storage by covering them with an oil or sterile distilled water and subculturing at regular intervals. The preservation of 'voucher specimens' in certain cases is advisable and may even be a legal requirement as, for example, in Commission Directive 93/85/EEC for control of potato ring rot in the European Community. Collections of stored material are an essential tool in diagnosis, by enabling comparison with authoritatively identified species, well-documented reference strains, or with previous isolations. Many countries maintain national collections of microorganisms for this and other purposes.

Accumulated diagnosis records build into national statistics and lists of pests recorded. Such statistics will inform the NPPO, plant health policy makers and administrators in planning, and in particular are useful in targeting inspections and monitoring operations on commodities and trades that present the greatest risk. Many countries publish lists of pests that have been recorded within their borders, often listed under their respective hosts and substrates. These are invaluable for many purposes, including PRA and in facilitating the rapid routine identification of pests for regulatory or advisory purposes. ISPM Nos 6 and 8 (Anon., 1997a, 1998a) provide guidance on the obligation to maintain pest records.

Pest Risk Analysis   11

Introduction

All plant health authorities must take decisions on matters that involve making a judgement on the immediate or potential threat of a pest to the well-being of crops and natural vegetation in the areas for which they are responsible. Many different examples could be given, but the following are some typical situations:

1. A previously unrecognized pest has been intercepted on imports and a decision on whether and what action to take is needed.
2. Information has been received that a pest of a domestic crop is spreading in the territory of a trading partner and a decision is needed on whether to introduce import controls.
3. An application has been received for initiation of a new trade in plant material, which has hitherto been banned and a decision is needed on whether to permit this and, if so, with what safeguards.
4. A country wishes to assemble arguments to justify or refute proposed national or international phytosanitary measures.

Not so long ago such decisions might have been taken subjectively by an experienced plant health scientist or administrator, perhaps after making some limited enquiries or discussion with colleagues. In effect, perhaps unconsciously, the plant health scientist or administrator would have made an unwritten PRA in coming to a conclusion. However, with such an individual approach, there was the possibility that the same data, when used by different people, might lead to very different conclusions on the risks posed by the pests or commodities in question, and thus to inconsistencies in the way in which pests were regulated.

Governments in most countries increasingly came under pressure to justify their decisions and actions with documented, consistent, logical and scientifically sound reasons, and a more uniform and structured PRA system was therefore sought.

The need to justify public expenditure and to provide a more demonstrable and solid base for decision-making prompted many early PRAs to concentrate on the estimated cost:benefit ratio of a proposed action, measure or eradication campaign (Pemberton, 1988). Arguments became more sophisticated during the late 1980s and an increasing number of factors were taken into account. Full PRAs started to become very complicated indeed, and needed to be done by highly qualified and experienced personnel with considerable inputs of time and resources, which was not always possible. A short, quick form of PRA was also needed, therefore, for less important decisions or to permit interim action while a full PRA was being done.

PRA has greatly increased in importance through requirements embodied in both the 1997 revision of the IPPC and the WTO-SPS (see Chapter 3 and ISPM No. 1, Anon., 1995). Amongst other things, these require that phytosanitary measures should be technically justifiable and that the rationale on which they are based shall be made available to other contracting parties on request. The science and practice of PRA has since developed substantially to provide such justification, and both the IPPC Secretariat and EPPO have published International Standards for PRA, which are complementary (ISPMs Nos 2 and 11, Anon., 1996b, 2001a; and EPPO PM 5/1, 5/2 and 5/3, EPPO, 1993a,b, 1997). These describe in detail the various steps that an assessor should follow in making a PRA, and up-to-date versions are available on the websites www.ippc.int/ and www.eppo.org/

The use of PRA as practised in the UK is described by Baker *et al.* (1999). This chapter is intended as a general guide to the principles of the subject and mentions some of the problems and difficulties that may be encountered. It does not attempt to provide practical instructions.

The Pest Risk Analysis Process

PRA is generally considered as having three main stages: initiation; pest risk assessment; and pest risk management (or what to do about the assessed risk). Much information has to be gathered before a PRA can be done and this process is described following an outline of the PRA process.

Initiation

The process begins with the initiation stage, identifying the reason for the PRA, the identity of the pest concerned and the geographical area under consideration. As mentioned above, there can be many reasons for initiating a PRA. The most common probably relate: (i) to an individual pest that has become significant in some way, for example, through an import interception, a new outbreak or spread to a new area; (ii) to a commodity that is being, or may soon be, imported; or (iii) to phytosanitary regulations that are thought to be needed or require revision.

Although the reasons may be varied, in practice the PRA will always be concerned with pests, either as a single species, as a group of species associated with a particular commodity, or as a group within a certain classification taxon, such as a genus or family. Sometimes assessment of pests may be necessary at a level below that of species, for example, at the level of sub-species, variety, strain or pathovar. However, pests are usually assessed individually at the level of species. At the start of a PRA it is therefore essential to be confident of the taxonomic identity of the pest to be evaluated.

A PRA relates to a particular area. This is designated the PRA Area and is normally a country, region or other geographical or political unit. The occurrence and regulatory position of the pest in the PRA Area will need to be established and related to its area of origin, history of spread and its present worldwide distribution.

Pest risk assessment: information needed

Information must be gathered and assembled to support the PRA. In doing this it may be helpful to prepare for each pest a datasheet containing all available information. In doing this, use of existing data sheets (such as those of Smith *et al.*, 1996) and information on the CAB International *Crop Protection Compendium* CD-ROM will be very helpful. The main categories of information required are as follows.

Pest identity

The identity of a pest is not always clear (Chapter 10). There may be controversy or confusion concerning the taxonomic classification of the organism. This sometimes makes it difficult to be sure that available information apparently relating to it is in fact reliable or relevant. This should be noted in the PRA, as well as any relationship to other known quarantine or regulated non-quarantine pests. Sometimes the identity of an organism may be reasonably clear although its taxonomy may not have been fully worked out or generally accepted. In these cases it may be acceptable to refer to a pest or host plant by a common name, such as (respectively) 'the cotton wilt *Fusarium*' or 'the florists' chrysanthemum'.

Pest distribution

The next step is to determine whether the pest already occurs within the PRA Area and, if so, its distribution. If this is not reliably known, it may be necessary to rely on such recorded observations as there may be. In the longer term, a survey may be needed to provide the information. It is also necessary to determine the pest's worldwide distribution, as far as it is known, which will help to identify the areas from which the pest might come and the pathways by which it might arrive. Many recorded observations may be dubious for some reason and may need investigation. For example, they may come from far outside the well-known range of the pest, or they may be suspected to be import interceptions rather than records from within a country.

Host plants

The known species of host plants for the pest must be identified and listed, together with their occurrence and distribution in the PRA Area. Polyphagous pests that have very wide host ranges may nevertheless show preferences for certain host species. Hosts may vary in susceptibility or sensitivity to attack and some may be attacked only when other, preferred, hosts are absent. Some pests, such as the heteroecious rust fungi and some aphids, may require more than one host in order to complete their life cycle, and the absence of one of them may negate the risk of the pest establishing. Cultivated varieties of the hosts found in the PRA Area and their susceptibility to the pest should be recorded. It is also vital to distinguish between hosts recorded as naturally affected and those recorded as attacked under experimental conditions. Otherwise a distorted picture may be obtained of the likely host range in the field. It is also possible that the PRA Area may contain plant species that, although not recorded as hosts, are nevertheless closely related to known hosts and therefore may be considered to be potential hosts.

Life cycle

The known characteristics of the pest's life cycle must be recorded. This should include a timescale and some measure of its progress, indicating the time needed for a new generation to be produced, how many generations there might be per year, and the relationship to the growing season and susceptible stage of the hosts. The need for sexual interaction is an important point in the life cycle. Many pests, such as aphids and fungi, can reproduce asexually apparently without limit. However, mating and meiotic cell division is necessary to produce plentiful genetic variation and the absence of a sex or a mating type will limit this, as shown by late blight of potatoes (*Phytophthora infestans*) in Europe before the introduction of the A2 mating type from Mexico in about 1976 (Shattock and Day, 1996). The mode of dispersal is another important feature of the life cycle and could greatly affect the conclusion of the PRA. The need for vectors and their occurrence in the PRA Area should be noted. Absence of essential vectors is an important point but, as with host plants, the presence of species closely related to known vectors must also be considered. Adaptability and survival in adverse conditions are other important pest characteristics. A life stage in soil tends to enhance survival, especially when combined with diapause or pupation in invertebrates, or with the formation of tough and long-lived resting structures in fungi or bacteria.

In considering the life cycle of the pest, information will be needed on the climatic and other environmental conditions in the PRA Area and in other parts of the pest's range of distribution, such as temperature, rainfall, soil type, etc. This is one of the most important aspects and it will be difficult to complete a satisfactory PRA without adequate and accurate information on these topics. Usually weather records are relatively easy to obtain as most countries maintain these, and soil maps or survey reports are frequently available.

Transport

A major consideration in PRA is the possibility of the pest being transported with traded plants and other commodities. For this the pattern of trade pathways connecting with the PRA Area must be determined, including the conditions (such as cold storage) and duration of transport and the state (such as dormancy or microplants) in which any host plants or plant parts are moved. For example, dormant plants without leaves will have a low risk of carrying leaf pests, while the risk of microplants *in vitro* harbouring pests is generally low, depending on the health of the parent plants and the system of production used. Records of interceptions of the pest or closely related species in trade, including occurrence on packaging or on the means of transport, must also be obtained.

Socio-economic impact

In assessing the potential socio-economic impact of the pest, any assessment that may already have been made of damage or yield losses due to the pest or near relatives will be valuable, even if the circumstances differ considerably from those in the PRA Area. In some countries, regular surveys and assessments are made of crop losses due to particular pests. If the pest has been subject to such assessment in any part of its range, this will be a source of valuable information. In any case, information will be needed on the type of damage caused on each major host, its seasonal variation and the economic damage thresholds, if these can be established. If climatic and cultural conditions in the PRA Area are not conducive to their expression, some types of damage may not be expressed, even if the pest and host both occur. Any routine phytosanitary measures and practices taken at the origins of the possible pathways should be investigated and noted, including any routine chemical or biological control programmes. The estimated costs of control if the pest becomes established will be needed to compare with the costs of exclusion or eradication. These costs must be estimated and may be affected by the interaction of control measures with those being used for the control of other pests, particularly if these include biological controls, or if the measures might have undesirable effects on the environment.

Pest risk assessment: handling information

PRA immediately generates a need for a large amount of information, not all of which may be available. The quality and value of the PRA depends very much on the quality of the information used in its construction. Ready access to reliable information is therefore of great importance. Although the amount of information available on-line through the Internet is steadily increasing, it is still essential at least to be able to do worldwide computer searches of abstract journals and databases, and preferably to have access to a good library or information centre. The PRA should carry full documentation for each stage. It should record the name and official position of the person making the evaluation, the reasoning involved, the date on which it was done, and the dates of information used. All sources should be acknowledged in the PRA, with detailed references listed in an orderly fashion. Where there are gaps in the information required, corresponding information on closely related organisms may be helpful to a certain extent, provided it is used and interpreted with caution and that this is made clear in the PRA. However, this will reduce the validity of the PRA and, if the gaps are too large, a meaningful PRA may be impossible to produce.

It is difficult to handle the large amounts of information successfully without the aid of a computer. Information may then be held on electronic databases, which can be accessed rapidly and the relevant data abstracted to support decisions, or for assembly into a PRA report. In many cases, to do this manually would be too laborious to be practicable, and it is notable that the development of PRA methods has taken place concurrently with the development of desk-top computers, their operating systems and programs. If resources permit, databases can be assembled on likely pests before an immediate need for PRA arises. This can be very helpful in assisting rapid assessment, but more often a lack of resources will not permit this approach. Cross-referencing between data is also helpful in facilitating the assembly of relevant information.

Pest risk assessment: evaluation

Before any evaluation begins, it is advisable to review the information readily available against the information that will be required, using the EPPO PM 5/1, 'Check-list of information required for pest risk analysis (PRA)' (EPPO, 1993a). If a previous PRA has been done on the relevant pest or one closely allied to it, this may form a valuable information source and starting point for the current PRA. The steps in evaluation are, first, to estimate the probability of the pest being introduced to the PRA Area; second, to estimate the probability of its establishment; and, third, in this eventuality, to estimate the economic impact. At the end of the process the risk is judged either to be acceptable or unacceptable, and in the latter situation the management of the risk is then considered and determined. In doing the first three stages, it is possible to allot numerical values in reply to various questions and then, as prompted in the EPPO system, to use these to calculate a figure for the magnitude of pest risk. The difficulty with this system is that not all the questions or replies will carry the same weight in relation to the assessment. Some aspects will be more important than others. Of course, it is possible to attach weights to the figures to counteract this, but such weights will largely be subjective in value and thus the final figure will also reflect this subjectivity. A numerical result in such circumstances may give a false impression of accuracy and lead to worthless comparisons being made between different figures. It is therefore important to be aware that numerical results and comparisons must be used only as a general guide and not regarded as being intrinsically objective and definitive. In some cases it may be best not to use numerical values at all, bearing in mind that the outcome will not necessarily be any more subjective than a numerical one.

In making an evaluation of the data, the PRA author may find it advisable and helpful to obtain specialist opinion on questions dealing with areas in which he or she is not expert.

Pest introduction

Although the FAO *Glossary of Phytosanitary Terms* (Anon., 2002a) defines 'introduction' as 'the entry of a pest resulting in its establishment', in PRA it is necessary to consider the probabilities of entry and establishment separately. The pathways by which the pest could enter the PRA Area from its existing geographical range should be listed and considered, including both natural and human agencies. As well as different forms of transport, entry via mail, packaging, dunnage, movement of people, animals and birds, and movement by wind or water should not be overlooked. The possibility of entry both alone and in association with hosts or other material should be considered. These pathways should be ranked in importance and individually assessed.

The likelihood of the pest getting into the pathway and surviving must also be considered. This raises the question of how probable it is that the pest would survive agricultural or commercial practices in the country of origin, or whether it might not be detected during phytosanitary procedures. Some life stages may be more risky (if the pest is very mobile) or more difficult to detect (perhaps within a substrate or as a latent infection), while some practices (such as kiln drying of timber) may be very effective in eliminating the pest. Survival in transit will depend on the speed and conditions of transport, the robustness of the life stages present and the number of individuals, spores or propagules involved. Obviously, the risk of entry will increase with the volume of host material carried by the pathway and with the degree of infestation.

Transfer of the pest from the pathway to a suitable host within the PRA Area will be more likely if the pathway distributes it widely or delivers it to favourable habitats (for example, where imported cut flowers are stored on nurseries producing ornamentals). The season during which pathway traffic occurs is important, as this may allow or prevent entry at times when the pest or host are at suitable life stages for establishment. For example, female invertebrates may enter when carrying fertile eggs, or there may be opportunity for male and female pests to meet and mate. The intended use of the host plants or commodity affects the chances of pest establishment. Consumption of these hosts raw, cooked or processed in other ways might reduce the chances of pest survival, while the production of waste in which it might survive, or the planting of host plants, would increase the chances of survival and establishment.

Pest establishment

In considering the pest's potential for establishment in the PRA Area, comparison should be made between the ecological and climatic conditions in the PRA Area and those prevailing in the various parts of the pest's distribution area. This must include consideration of microclimatic conditions in protected cropping and their possible correspondence with ambient conditions in warmer climates, or vice versa. For these purposes geographical information systems (see below) may be a helpful tool.

The abundance of the host plants and (if necessary) vectors in the PRA Area will greatly affect the risk of pest establishment. Usually the presence of substantial populations of host plants and any necessary vectors in the PRA Area is essential for successful establishment. Where a pest occurs but lacks an efficient vector, the PRA may need to concentrate on the vector species rather than the pest itself. This risk was forcefully illustrated in the UK when, in 1986, the introduction of an efficient vector, western flower thrips (*Frankliniella occidentalis*), resulted in widespread damage to glasshouse ornamentals from dissemination of *Tomato spotted wilt virus* (Baker et al., 1993). If hosts of crop pests are present in the natural vegetation, or could be present as crop weeds or volunteer plants, this may increase the risk of establishment. Other biological factors that could affect the likelihood of pest establishment include the presence or absence of natural enemies of the pest and the pest's adaptability as shown in other parts of its range. Non-biological factors include soil type, topography, environmental pollution, agronomic factors (such as the type and method of cultivation and other operations), use of irrigation and the methods of pest control already in use. Control measures used against the pest in any part of its range and their success should be detailed, and the actual or possible development of resistance to pesticides should be noted. If the pest is already known to have been introduced in new areas, this will give a good indication of its ability to overcome unfavourable factors and the risk of its establishment in the PRA Area may be considered to be greater. However, if the pest already occurs but is limited by environmental conditions, its potential for further spread may be small. The possibility and ease of eradicating or containing the pest, if it should eventually become established, should also be considered.

Socio-economic impact

Consideration of socio-economic losses from the pest in other parts of its global range is a suitable starting point for evaluating its potential socio-economic impact in the PRA Area. Crop losses of yield or quality, either in the field or under protection, are relatively easy to cost from commodity prices, but environmental damage and loss of amenity can be much more difficult to quantify in monetary terms, even if the damage is obvious. In some situations, even small crop losses may make cultivation of the host crop uneconomic, or may shift the export/import balance for a commodity. It is also possible that the pest may cause economic damage, not so much by direct losses of yield as by its effect on the eligibility for marketing or export of its hosts or host-derived commodities. It can happen that ecological conditions in the PRA Area may be unfavourable for significant economic damage to occur, even though they may permit the pest to establish. Impact on individual growers or traders may be more serious than impact on the national economy. It is possible that not all the PRA Area may be threatened by the pest. If so, then the *endangered area* should be defined and its extent considered. The socio-economic impact will also be affected by the rapidity with which the pest could spread if it becomes established. Obviously, a potential for only slow spread (as with many soil-borne pests) would have a more gradual impact than rapid spread. The economic effect of the pest on the production system and consumer demand must also be considered. The need to control a new pest will usually result in increased production costs. Consequential increases in commodity prices or decrease in quality may reduce consumer demand and result in poorer sales. The presence of the pest may also have other, indirect, undesirable economic effects, especially if control with chemical pesticides might be necessary.

Final evaluation

An important consideration at this point is whether the pest is an already known and recognized quarantine organism or serious pest elsewhere, or whether it is a newly recognized hazard. The known presence or absence of the pest in the PRA Area is also very important. If it appears that the pest is very likely to spread to the PRA Area by natural means in the near future, this may affect the justification of imposing restrictive measures to stem its spread in trade. After all the available relevant information has been assembled and considered, the PRA assessor will have to make a judgement on the overall risk that the pest presents, and on whether and what measures should be taken to prevent its establishment. In doing this, the quality and adequacy of the information available should be mentioned in the PRA report and taken into account.

Pest risk management

Bearing in mind the level of phytosanitary protection required and that this level should be appropriate to the risk, phytosanitary measures may be proposed, which are justified by the PRA. These may be separated into short-term and longer-term measures, and will usually involve some of those discussed in Chapter 7. Where the introduction of phytosanitary measures is subject to national or international approval, the PRA will also provide the basis for argument as to the necessity and justification for the measures proposed.

Preliminary Pesk Risk Analysis

It is often necessary to make a rapid and necessarily abridged preliminary PRA on which to base a decision for immediate action. This may be needed because a pest has been intercepted on perishable produce that cannot be held for long without impairing its marketability, or perhaps because it has been found in an active state in the field and immediate action must be taken if it is to be eliminated and establishment prevented. Such a preliminary PRA can use only information that is readily available. If the PRA indicates that the pest could be a serious threat, it will usually be prudent to take a precautionary approach (Griffin, 2000), especially if important information is unavailable. Such an approach is supported by the IPPC (Article VII.6) and the WTO-SPS Agreement (Article 5.7). In this case a full PRA should be done as soon as possible and any stringent measures already taken must be relaxed if the more complete PRA later reveals the risk to be lower than it at first seemed, or if less stringent measures would achieve the necessary level of protection. Conversely, the preliminary PRA may indicate that the pest is unlikely to be of significance and that it is not worthwhile to pursue further assessment. In this case, no phytosanitary measures should be imposed. It can be prudent to do a preliminary PRA even if there is no great urgency for a result. A guide to such a simplified PRA has been published by EPPO as PM 5/2 (EPPO, 1993b) and the practical use of such a PRA in the UK is shown by Sansford (1998) for Karnal bunt (*Tilletia indica*) of wheat.

Geographical Information Systems

When assembling and assessing information for PRA, much of the data handled concerns the distribution of pests and hosts, the variation in climate from place to place and other data which can be illustrated on geographical maps. In recent years computer programs (such as CLIMEX) have been developed to handle and analyse such data and display them in map form. These are known as geographical information systems (GIS) and greatly facilitate assessments in PRA when large amounts of mappable data need to be processed and compared, for example, when using weather records to compare climates in different locations.

Climate mapping

Techniques for climate mapping are described by Baker *et al.* (2000). The CLIMEX program contains world meteorological data for 1931–1960, which can be used in comparing climates worldwide. If desired, additional meteorological data can be imported to augment this, when available. However, it often happens that weather stations are located in places such as urban areas, on coasts, or at airports, which may not be representative of the PRA Area or of other relevant areas, such as those of a pest's current distribution. Provided that reliable digital map data are available for the areas in question, this problem can be ameliorated by interpolating the weather data for the areas between weather stations, making allowance for factors such as latitude, topographical elevation, aspect and proximity to the sea. For this purpose, the interpolations are usually related to map grid squares or area units bounded by degrees of latitude and longitude. These can then be presented as a map showing how the climate changes over the landscape; for example, to show areas more or less suitable for pest survival. This interpolation process may have to be done before importing the data into the GIS program. The resulting maps can be displayed at any scale up to the largest permitted by the digital map data available. If the scale is large enough to show individual holdings, fields and crops, the maps can illustrate those that may be more or less suitable for pest development, the estimated speed of development in different places, the number of generations per growing season, and many other aspects of pest biology. In turn, this will assist with prioritizing fields for survey, estimating likely areas for pest survival, and many other aspects of PRA and pest control management, especially if used with current season weather data. The program can also show the effects on pest biology of introducing theoretical factors, such as climate change of varying degree.

Climographs

Frequently the environmental variables affecting the distribution and life cycle of the pest being assessed are not well known. In the longer term some of these can be determined by experimental research. Meanwhile the climatic variables can be estimated from study of meteorological data covering areas of the pest's existing distribution and known abundance. The climates in which the pest is abundant, occasionally abundant, rare or absent can be compared with those in the PRA Area to indicate the likelihood of the pest being able to establish and, if so, its likely abundance. For each of these climates, data for different significant climatic variables can be plotted against each other to produce climographs. These graphs facilitate comparison of the climates and estimation of the pest's potential abundance in the PRA Area, as illustrated in Fig. 11.1.

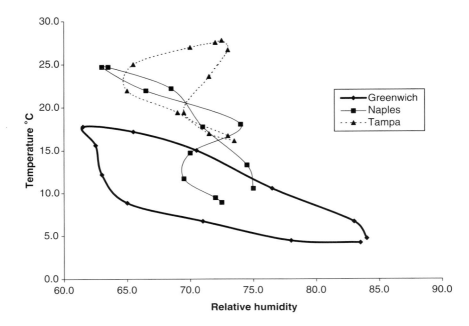

Fig. 11.1. Climograph for the Mediterranean fruit fly, *Ceratitis capitata*, based on mean monthly climate data, 1931–1960. Data for consecutive months from January to December are connected with lines. The similarity between the parameters for Naples, Italy (where *C. capitata* occurs) and Tampa, Florida, and their difference from Greenwich, UK (where *C. capitata* does not occur) contributes support to a conclusion that *C. capitata* might have the potential to establish in the climate of Tampa. (Courtesy of Dr R.H.A. Baker, CSL.)

Indices

The suitability of an area for pest establishment, or the variation in abundance of a pest, is often determined by many interrelated factors. If key factors can be determined, these can be combined mathematically in various empirical ways to produce an index, which can be used to compare the suitability of areas for pest establishment and potential pest abundance. The use of indices for comparing climates in assessing the risk of establishment of Karnal bunt (*Tilletia indica*) is demonstrated by Sansford (1998). The indices developed must be related to area units and validated by comparison with the known distribution of the pest. When found to agree well with the pest's existing distribution, the relevant index can then be mapped on the PRA Area to indicate the likely potential of the pest, as shown in Fig. 11.2.

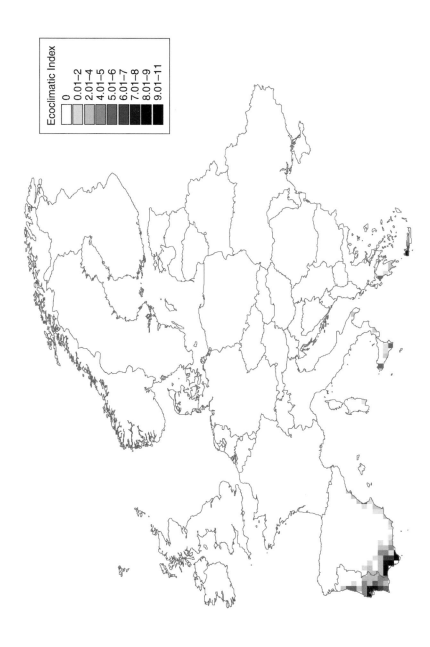

Fig. 11.2. The establishment potential of the Queensland fruit fly, *Bactrocera tryoni*, in the Mediterranean area, estimated by an ecoclimatic index calculated with the CLIMEX computer program. The University of East Anglia Climatic Research Unit climatology, which interpolates climates to a 0.5° latitude × 0.5° longitude grid has been used. CLIMEX outputs have been mapped using a geographical information system. Darker shading indicates greater potential for establishment. (Courtesy of Dr R.H.A. Baker, CSL.)

Phenology models

Phenology models are used to predict the timing of events in an organism's life cycle. For this the relationship between the progress of events in the life cycle and critical climatic factors, such as temperature, must be known. Complete daily sets of meteorological data relevant to the PRA Areas must also be available, and interpolation of the data from existing weather stations is necessary to estimate the variation of the data over the landscape. These data can then be used to display with maps the way in which the pest's life cycle is likely to vary at different places over the landscape. Phenology models can also be used to study aspects such as the potential distribution or establishment of a pest, where these are dependent on the pest or host reaching a particular life stage. For example, it may be necessary for a pest to reach a robust stage of its life cycle in order to survive an unfavourable period, such as winter or a dry season. A good example of the importance of studying the relationship between host and pathogen phenology is the assessment of the risk of establishment of Karnal bunt (*Tilletia indica*) of wheat in the UK and continental Europe. Sansford (1998) showed that environmental conditions conducive to infection existed in both India and the UK at the time of ear emergence, when wheat is susceptible to infection. Study of probable pest development using meteorological records from extreme years can reveal the potential limits to the risk. A pest might establish or survive in exceptionally favourable years, or it might die out in exceptionally unfavourable ones.

In relating daily weather data to actual dates and handling these in computations, it is necessary to use a numerical figure for calendar dates. For this it is usual to use the Julian day number (J), which is the number of the day determined by counting sequentially from day 0, starting at noon on 1 January 4713 BC (the first day of the Julian cycle introduced by Joseph Scaliger in 1583). This number is normally generated by the GIS program used.

The value of geographical information systems

As with numerical PRA values, the strong visual representations of pest risk provided by a GIS program may give a misleading impression of reliability and accuracy. When interpreting GIS-produced maps it should be kept in mind that many factors other than those taken into account by GIS may affect pest development and establishment.

Prediction and estimation of pest establishment, abundance and dispersal by geographical mapping and comparison of climates has been criticized for not taking into account the effects of competition between species occupying the same or similar ecological niches and of natural enemies. However, it will seldom be possible for all the multitudinous factors affecting a pest's potential distribution to be taken fully into account because of the lack of appropriate data. Estimates must be made with the data and labour resources available. Risk assessments using GIS can take account of a great many factors when information is available, but it is likely that in many circumstances interspecific competition and the effect of natural enemies may be less important than other factors, including non-climatic ones. GIS remains an important and valuable tool in the practice of PRA.

Hygiene and Precautionary Measures   12

Good Plant Health Practice for Growers and Other Businesses

As in all disciplines concerned with the handling of plants, plant material, plant products or various kinds of plant pests or pest situations and preventing their unintentional spread, good hygiene is central to successful plant health operations. However, there are many other related practices that also contribute to success. Most of these are simple and common-sense precautions, which should be part of normal routine. Although the basic practices are common to all plant health situations, some naturally vary according to the type of business or operation.

Personnel

In each business that deals with plants, plant materials or plant products, especially if any are subject to phytosanitary controls, it is helpful if one senior member of staff is designated as having overall responsibility for plant health matters. This person should have at least basic training in the biology and management of plant pests and should be the first point of official contact on all plant health matters. Good liaison should be maintained with local phytosanitary inspectors or other local representatives of the national plant protection organization (NPPO).

Sources of planting or propagating material

Whether sowing a field crop, planting a new orchard, or planting up a new plant nursery propagation bed, good quality starting material is a key requirement for success. In each case the starting material needs to be correctly named, vigorous, and as free from mixtures and healthy as is necessary for its purpose. For the nursery or farm it may be possible to obtain good planting material for appropriate crops from a recognized certification authority for plants or seed. To a substantial degree, such material will carry assurance that it is of the specified species, subspecies and variety, is not mixed with other species to an extent that would impair its usefulness, and that it is sufficiently healthy and vigorous for production purposes (Chapter 8). In addition, such material will have a known provenance and, in case of any fault, its ancestry will be traceable.

Field crop hygiene

Weeds frequently harbour pests, provide them with alternative hosts, or make pests more difficult to detect. Apart from reasons of preventing competition with crop plants, minimizing the growth of weeds is therefore a key practice in plant health as well as good crop husbandry.

Avoiding growing successive crops that share susceptibility to the same soil-borne pests on the same piece of land prevents the establishment and accumulation of these pests in the soil. Such crop rotation is a basic precaution against the build-up of pests in the growing medium or soil, and rotations of various lengths of time are necessary to keep various different pests in check. Although rotation can be dispensed with for some crops which are not subject to serious soil-borne pests, or for which such pests can be otherwise controlled, nevertheless, in general it is good practice. This applies to the garden or allotment plot as well as to large arable farms, minimizing losses from soil-borne pests and reducing the need for pesticide applications.

Pests frequently enter crops via the water used for irrigation. It is therefore most important to ensure that the source of the water used is free from pest organisms. Water from a public or commercial authority supply or from a deep borehole is normally safe in this respect. However, more caution is needed if the source is a reservoir on the farm or holding or a watercourse, either natural or artificial. In these cases careful investigation should be made to ascertain whether there is any risk from possibly contaminated crops or alternative hosts within the reservoir catchment area or upstream of the abstraction point. Water-borne pests can travel considerable distances (a kilometre or more), depending on the infection source and degree of water movement. Well-documented cases have traced such movement with pests such as *Ralstonia solanacearum* on *Solanum* potatoes (Elphinstone *et al.*, 1998), *Phytophthora fragariae* on strawberries, and *Beet necrotic yellow vein virus* (the cause of rhizomania disease) of sugarbeet vectored by *Polymyxa betae*.

Disposal of field crop wastes

Careful disposal of crop wastes and minimizing the movement of soil are good basic plant health practices. Burning of crop wastes is not always either possible or desirable. Burial by prompt ploughing or removal by foraging animals may be satisfactory. For example, pigs efficiently remove ground-keeper potato tubers, while sugarbeet tops and residues can be grazed by sheep. Particular care is needed where there is a risk of carry-over of pests from one crop to another. For example, inoculum of potato late blight (*Phytophthora infestans*) may persist in tubers discarded or remaining from heaps or clamps made at harvest time. Infected shoots arising from these the following season represent an important source of infection for potato crops in the vicinity. Wastes (including soil) from plant produce cleaned, packaged or processed on the farm should be returned to the field of origin wherever possible.

Where serious soil-borne pests occur, it is common practice to place a wheel bath (usually containing a disinfectant) at the entrance to the farm or area, with the aim of minimizing the movement of soil on to or off the premises. While this may serve to draw attention to the need for hygiene precautions, it is doubtful that it is particularly effective, as it will not remove or penetrate much of the soil adhering to the vehicles passing through it and the concentration of disinfectant will rapidly decline if not regularly replenished.

Crop hygiene in protected environments

Protected environments can be more conducive to the rapid increase of plant pests. They can be particularly prone to the establishment of exotic pests. Protected crops are therefore usually more vulnerable to pest attack than field crops. Good hygiene in protected environments is therefore extremely important to successful production. The basic principles are the same as described above for field crops, and cleaning and disinfection in protected environments are discussed below. However, particular care should be taken not to bring plants and plant materials of unknown plant health status into protective structures housing crops vulnerable to shared pests. This could be especially risky if the material is imported, and many protected crops have been lost to pests introduced in this way, for example, through the storage of imported ornamentals or cut flowers in structures housing other susceptible crops. It is best to use separate sets of implements for different units, or at least to clean tools when moving from one crop or unit to another. This is particularly so for knives and other cutting implements, which should also be disinfected. It is often convenient to use disposable razors when taking cuttings. Cured and processed tobacco can still harbour viable viruses (such as *Tobacco mosaic virus*) to which the tobacco plant is susceptible. Staff working on protected crops should therefore avoid smoking, especially when dealing with solanaceous crops. They should also wash their hands and change protective clothing when moving between units containing crops of different health status.

Cleaning and disinfection

Although not normally part of farm and field husbandry, cleaning and disinfection are an integral part of plant production within protective structures such as glasshouses and plastic-covered tunnels. Where these have an impervious floor, this should be kept clean and free from debris by regular sweeping and power washing. Likewise, containers used for plant growth should be cleaned and washed after use and before the next crop is started. Any crop debris and used growing media should be collected and removed to a covered pit or other container. It can then be disposed of or treated by composting, steaming, fumigation, incineration or removal to an approved waste disposal site. Plants and growing media known to be contaminated by important pests should be promptly removed, placed in sealed bags and incinerated or disposed of by deep burial at an approved waste disposal site. Premises that are kept tidy and orderly are easier to keep clean and free from contamination.

For protected crops produced within permanent structures, it is good practice, where possible, to arrange cropping so that there is a complete break in cropping at one point in the season. This permits a general clean-up of the premises to be made and chemical or other treatments can be used, which would not be possible if growing plants were present. Where appropriate, the structure and growing medium can then be treated with suitable chemicals to kill contaminating microbial pests or the eggs, larvae or adults of invertebrate pests. Structures may be washed and cleaned with high-pressure water or steam hoses or treated by spraying with chemical pesticides or disinfectants. Space treatments to kill harmful organisms in the air, which may also be effective in disinfecting the interior of structures, are normally applied as mists, fogs or smokes (according to the size of the droplets or particles generated), or as fumigant gases.

Growing media or soil can be disinfested by applying steam, either *in situ* or in pits or containers. This method has been used in horticultural practice for well over 100 years. The surface must be covered to ensure sufficient heat is retained. The steam may be introduced via permanent pipes or through the sides and base of the container and must be applied for sufficient time for the required temperature (or higher) to be reached throughout the bulk for the period of time necessary to kill relevant harmful organisms (see below). However, as with some forms of composting, material at the edges or periphery of the bulk may not reach the temperature required, and to be fully effective the material may have to be mixed and the steam reapplied. Depending on the chemical used, fumigation is usually more effective because the fumigant gas penetrates more easily to all parts of a bulk, provided it is within a sealed container and the temperature is warm enough for the fumigant used. Soil can be fumigated effectively to cultivated depths either indoors or outdoors, provided it is covered with an impervious material (usually polyethylene sheet) and the soil temperature is warm enough. Depending on the fumigant used, the material being treated and the harmful organisms to be killed, there are specific time and temperature requirements that must be fulfilled for effective treatment. At normal temperatures these times usually range from several hours to days. Operator protection and other safety practices in accordance with manufacturers' recommendations and national legislation must be strictly observed. As most fumigants are extremely toxic to both plants and animals, the fumigant must be allowed to dissipate after treatment and this may take a week or more before the soil or growing medium is safe for planting.

Pesticides

Most countries regulate the use of pesticides and in many there is an extensive suite of legislation and regulatory measures governing their marketing and application. In Member States of the European Union, for example, there are complicated and comprehensive laws governing pesticide approval, registration and application, including requirements for efficacy, environmental toxicology, formulation, naming, packaging, labelling, minimum periods between application and harvest, minimum residue levels in various crops, sprayer operator qualifications, and many other aspects. It follows that any pesticides routinely used should be those approved for the purpose and applied in the prescribed manner. In particular, any specified safety precautions should be strictly observed. Stocks of outdated or withdrawn pesticides should be disposed of in a proper manner. When pesticides are withdrawn there is often a period during which they may continue to be used, and disposal by application in the normal way during this period is usually the best method. Otherwise they should be disposed of to an approved toxic waste site.

Even where 'organic' farming is not being practised, the general aim should be to minimize the use of pesticides. EPPO has drawn up guidelines on Good Plant Protection Practice for many crops, which explain the agreed best strategy for controlling important pests with minimum pesticide application. In many cases this can be achieved by the use of pest assessment or meteorologically based forecasts and knowledge of particular infection periods for various pathogens, or by use of an integrated programme incorporating biological control methods.

Storage and transport

As the pest status of home-grown produce will usually be known and under the control of the producer, it will normally carry a lower phytosanitary risk than material of unknown pest status from elsewhere. Therefore, it is advisable to handle and store home-grown material separately from that brought from elsewhere, especially if it is imported from abroad. In particular, waste from brought-in material should be stored and handled separately. Waste should be stored in pits or containers that do not allow seepage into production land, drains or watercourses. Waste stores should also have covers that prevent waste from being removed by animals, birds or the wind.

All vehicles and equipment used for waste transport should be cleaned and, if necessary, disinfected after use and before using them for other purposes. Brushing or washing with a pressure hose should be sufficient to remove adhering soil and debris. Bags or sacks used for transporting waste should be sealed and disposed of together with the waste.

Record keeping

This is an aspect of good plant health practice that is often neglected, especially where not legally required for some particular purpose, but which is important for the efficient running of all businesses concerned with plants. For nurseries and farms an accurate large-scale map of the holding is a prime requisite. This is essential for recording the location of various crops from season to season, the site of any outbreaks of soil-borne pests, and any treatments applied to the land. Keeping detailed records of material bought and sold, suppliers and customers, occurrence of significant pests and action taken, dates and rates of routine or emergency treatments applied, dates of planting, cultivations and harvesting, and an up-to-date inventory of plant material held should be routine practice. This not only facilitates the efficient running of a business, but when problems arise it also permits trace-back to identify the causes. For larger businesses, computerized record keeping can reduce the workload of this task, provided the system is not so complicated or unreliable that it causes problems of its own. Records need not be retained for very long periods; retention for about 2 years would satisfy normal requirements. However, those relating to crop rotations should be kept for at least the length of one crop rotation cycle and preferably for several cycles, especially if these are of short duration.

Additional Hygiene Precautions for NPPOs and Containment Facility Operators

Protective clothing

Overclothes, which are disposable, or which can be cleaned and sterilized easily, are important in minimizing the risk of spreading pest organisms in many plant health situations. Rubber boots reaching to below the knee ('Wellington' boots) are required for many situations in the field, nursery or glasshouse where soil, liquids or other contaminated material may adhere to footwear. Such boots can be washed clean with water and treated with a suitable disinfectant to prevent spreading contamination. When cleaning boots, close attention is needed to remove soil or contaminants from the cleats on the soles, which otherwise can retain soil through even vigorous washing. Boots that do not have large cleats are therefore preferable. Rubber boots can be supplemented with waterproof over-trousers and jackets, which can be washed clean with a brush or jet of water after visiting contaminated areas or pushing through infected or infested vegetation. Disposable overalls, over-shoes, and hair covers are often useful in maintaining hygiene when working in enclosed areas containing pests such as thrips, or fungi, which produce air-borne spores.

Disinfectants and fumigants

Disinfectants

The term 'disinfectant' is usually taken to apply to those pesticides active by contact against microbial pests, including viruses. It commonly refers to chemicals used in medical and veterinary practice to kill and prevent the spread of human and animal pathogens. However, disinfectants are often needed in plant health work, both in routine situations such as crop inspections and for use in eliminating infective agents when combating outbreaks of microbial plant pests, or in laboratory work. For example, phytosanitary inspectors may need to disinfect boots or protective clothing after inspecting potentially infected crops, and saws, secateurs, knives and other implements may need disinfecting after pruning to control diseases such as fireblight (*Erwinia amylovora*) on pome fruit trees and related ornamentals. Disinfectants are also used in plant health research, as in other areas of plant pathology, where it is necessary to disinfect laboratory or glasshouse premises, or to surface sterilize plant material before isolations are made. However, many disinfectants that are effective against air- or water-borne pathogens are much less active against soil-borne pathogens, which often have life stages with tough outer cell walls or are situated within protective organic material.

Ideally, it would be desirable to use a single disinfectant for all plant health field work, to avoid the need for phytosanitary inspectors to carry different disinfectants for different tasks. Where there are close working links to animal health, a single disinfectant effective against animal pathogens as well as plant pests would also be desirable. However, disinfectants are not equally effective against all types of pests. For example, those effective against bacteria may not be effective against viruses or fungi, and vice versa. This creates difficulties where phytosanitary inspectors have to carry different kinds of disinfectant to deal with several different types of pest, and efforts have been made to find a disinfectant with as wide a spectrum of antimicrobial activity as possible. There are also other criteria that disinfectants should satisfy. They should not be corrosive or otherwise dangerous to handle, they should be easy to dilute or prepare for use, and they should not be unacceptably unpleasant to use (for example, in smell, consistency or persistence). No single disinfectant meets all these desirable criteria, but some are effective against more than one type of pest and are otherwise acceptable to use. There are many proprietary disinfectants on the market that have activity against different ranges of pests. In general, these should be used only against the types of pests for which the manufacturers claim they are effective and the manufacturers' prescribed rates of application and safety precautions should be followed closely. However, the spectrum of activity claimed by manufacturers does not always agree with what is found in plant health practice. Therefore, especially where there has been no previous experience in the use of a disinfectant under local conditions or for a particular purpose, it is advisable to carry out preliminary trials to confirm its efficacy before using it for plant health purposes.

A very large number of chemicals have disinfectant activity and many proprietary disinfectants consist of mixtures of several different chemicals. Disinfectant chemicals are of many different types. They are often powerful oxidants, organic acids, phenols or chlorine compounds. Many effective disinfectants are relatively simple compounds. For example, sodium hypochlorite and ethyl alcohol are widely used for surface sterilizing plant material, and trisodium orthophosphate is used in the UK as a general disinfectant against plant viruses. Formalin, a widely marketed aqueous solution containing about 40% formaldehyde, is a very effective disinfectant for many plant health purposes, and has been so used for very many years. However, it is unpleasant to use in confined spaces and, like lysol, can also be carcinogenic. Its use is therefore to be avoided where possible and alternatives have been sought (e.g. Reed and Dickens, 1993). Effective alternative materials are available for use in many circumstances, but none have such a wide spectrum of activity.

Fumigants

Fumigants are gaseous pesticides and have a much greater capacity to penetrate materials such as soil, tightly packed plants and bulk stores of grain, than do solid or liquid pesticides. Fumigants may be applied as liquids or solids as well as directly as gases. Liquid materials are generally either gases liquified under pressure, gases dissolved in a solvent, or highly volatile compounds. Parent materials give rise to the active gaseous pesticide by volatilization, chemical decomposition (usually involving reaction with ambient moisture) or some other process. There is normally a temperature threshold below which the fumigant gas is not effective or not generated in sufficient quantity. Effective dosage rates depend on many variables, including soil type, porosity, moisture content, temperature and duration of treatment. Time/temperature combinations must therefore be established for each fumigant and material to be treated.

In plant health work, fumigants are much used for routine pre-shipment, emergency and hygiene treatments. They are particularly useful for space treatments to eliminate pests in enclosed environments such as glasshouses, transport containers and the holds of ships (see Chapter 7). Material to be fumigated, which is not already enclosed, must be placed in a fumigation chamber or specially constructed fumigation tent before treatment.

SOIL FUMIGATION. Fumigants are also extensively used for treatment of soil, both as an emergency phytosanitary treatment to eliminate outbreaks of important soil-borne pests and as a routine horticultural practice to ensure freedom from soil-borne pests before planting high-value crops such as strawberries. Usually the soil must be covered with impervious material before or after fumigant application, or at least the soil surface must be compacted by rolling, to retain the fumigant for a sufficient length of time. Polyethylene sheeting has been commonly employed for this purpose, but in many countries (including the European Union, under Regulation 2037/2000) virtually impermeable film sheeting must now be used. Liquid and gaseous fumigants are applied to the soil by injection, while solid materials are applied as powders or granules. The most effective soil fumigants are usually highly toxic to most living organisms, including plants, so are normally applied to bare land and allowed to disperse before the crop is planted. Soil fumigation in the field does not completely eliminate all the soil fauna and flora, although it may eliminate the target pest. It may also affect the subsequent crop by inducing the release of plant nutrients that are already present in non-available forms, direct provision of plant nutrients, and by altering the balance of soil fauna and flora populations.

M. le Baron Paul Thenard was the first to use soil fumigation to control a plant pest (Wilhelm, 1966). In 1869 he treated the soil of a vineyard near Bordeaux, France, with carbon disulphide and found that this gave good control of *Viteus vitifolii* (phylloxera of the vine). This initiated research on, and development of, soil fumigation in many countries, and numerous chemicals for soil fumigation are now on the market. Common soil fumigants include chlorinated hydrocarbons, organic thiocyanates, formaldehyde and methyl bromide. The liquid fumigant DD, for example, consists mainly of a mixture of 1,3-dichloropropene and 1,2-dichloropropane. A by-product of alkyl plastic manufacture, it was introduced as a soil fumigant in the USA by W. Carter in 1943 and is mainly used as a nematicide. Several fumigants rely on the production of methyl isothiocyanate for their activity. Amongst these is dazomet, which is applied in powder form and is mainly dimethyl tetrahydrothiadiazine thione. In the soil about 80% of the compound decomposes under catalysis of clay particles and moisture to yield monomethylamine, formaldehyde, hydrogen sulphide and carbon dioxide. It is effective against a wide range of pests, including bacteria, fungi and nematodes.

METHYL BROMIDE. Up to now, the material most widely used for fumigation of many types of plant material, plant products and soil has been methyl bromide. Being a gas at temperatures above its boiling point of 3.6°C, it is handled as a liquid under pressure and is applied by injection. It is effective against a very wide range of pests, including invertebrates, fungi and bacteria, but is not phytotoxic to many plants when applied at appropriate rates. It is effective at relatively low temperatures, is fast acting, leaves harmless residues, and pests do not develop resistance to it. Methyl bromide is the basis of a wide range of phytosanitary treatments against many different types of pest, which have been published by the FAO and EPPO, and in many cases complete control of the pest can be achieved. However, methyl bromide has been identified as an important member of the group of chlorine and bromine compounds that deplete stratospheric ozone and so allow greater amounts of harmful solar UV radiation to reach the Earth's surface. For this reason its future production and use is to be severely curtailed and controlled.

The Montreal Protocol. In response to concerns that the Earth's ozone layer was being depleted, the United Nations Environment Programme established, in 1981, a working group to promote a treaty aimed at preventing this. After considerable difficulty, The Vienna Convention for the Protection of the Ozone Layer was agreed in 1985. Participating nations agreed to take appropriate measures 'against adverse effects resulting ... from human activities which modify or are likely to modify the Ozone Layer'. The Convention provided for future protocols on specific action and set out procedures for amendment and dispute settlement. Almost immediately, research results were published, which showed severe depletion of the Ozone Layer over the Antarctic. This provided impetus for seeking agreement on specific measures, and the Montreal Protocol on Substances that Deplete the Ozone Layer was signed in September 1987 and came into effect in 1989. The Montreal Protocol controls the production and consumption of substances that can cause ozone depletion. These include the chloro/fluoro-carbons and many other halogenated compounds, including methyl bromide. In 1990, the participating governments agreed to phase out production and use of controlled substances on a rather lengthy timescale, but in subsequent meetings the schedules for this phasing out were accelerated. The Protocol is subject to amendment and the schedules distinguish between developed and developing nations. The current general phase-out schedule in developed nations required reductions of 50% in production and importation relative to the 1986 baseline by 2001 and complete phase-out in 2005. For developing countries consumption was to be frozen at 1995–1998 average levels by 2002, followed by complete phase-out by 2015. However, there is provision for continued production and use after these dates of limited amounts of controlled substances essential for health and safety and for which there are no technically, economically or environmentally feasible alternatives available. Such Critical Use Exemptions must be applied for and will be evaluated by the Secretariat and the Technology and Economic Assessment Panel of the Montreal Protocol, and would cover many uses of methyl bromide for plant health purposes.

Up-to-date information is available at the websites www.unep.org/ozone/vienna.shtml, www.ars.usda.gov/is/mb/mebrweb.htm, www.epa.gov/ozone/mbr and www.ea.gov.au/atmosphere/ozone/

Methyl bromide in the atmosphere substantially derives from the metabolism of marine algae as well as human activity. The relative contributions of these sources to atmospheric methyl bromide have not been firmly established, but anthropogenic emissions are thought to account for about 25% of the total. Of this, about 75% derives from soil fumigation and less than 20% from the use of methyl bromide for post-harvest fumigation and phytosanitary treatments. Alternatives to methyl bromide are available for many purposes, including partial soil sterilization and phytosanitary treatments of plant products. However, no single alternative to methyl bromide has its wide range of applications and, in comparison, most substitutes have other and substantial disadvantages. For example, phosphine is not effective for soil fumigation and it requires much longer periods of application than does methyl bromide, but it does have potential for fumigating live plants when gaseous formulations are used. Depending on the circumstances, partial soil sterilization can be achieved by steaming or (in warm climates) soil solarization (Katan, 1981) and there are other soil fumigants available (e.g. chloropicrin, formalin). However, all have disadvantages, either in efficacy or in requiring higher soil temperatures or longer post-treatment periods, and some may leave potentially harmful residues in the soil. Numerous other phytosanitary treatments are available for different circumstances, as described in Chapter 7.

Waste disposal

Waste disposal is an important aspect of phytosanitary hygiene and in many countries is covered by legislation or official guidelines, as in the UK (Anon., 1998c). Waste material that is of phytosanitary concern may be liquid or solid and may derive from many sources. For example, it may come from commercial activities such as washing, trimming, peeling, grading and packing plants or plant produce on farms or in factories, at large-scale cooking premises, or during storage and transport operations. Alternatively, it may come from private or public scientific research or activities related to testing of plants, plant material or soil for pesticides, pests, or other characters, or developing new pesticides or plant varieties. The waste may consist of soil or other growing media, peelings, discards and other plant debris, or wash water. Large quantities of soil or plant debris may accumulate at factories processing sugarbeet, sugarcane, potatoes or other produce, and this may present considerable difficulties in disposal. Very often it is returned to agricultural land and is thus a potential means of dispersal for any viable pest it may carry. Similarly, such factories use large quantities of water and, after use, this may be discharged to watercourses. Any viable water-borne pests in such waste could be recirculated to crops by abstraction of water for irrigation further downstream.

The type of waste generated and its potential phytosanitary risks must be assessed before a satisfactory waste management policy can be decided. For example, waste from imported material might generally be regarded as carrying a greater phytosanitary risk than that from domestic produce, but there may be areas or types of crop in which serious pests occur in domestic production and for which the phytosanitary risk would be high. Soil-borne pests are often persistent and remain viable for many years, so waste soil is usually of greater risk than plant waste. In many cases the type of commodity from which the waste comes, its range of potential pests and its national importance will suggest whether it is of high or low risk. For example, where *Solanum* potatoes are important in national agriculture, the phytosanitary risk of waste from imports deriving from areas where serious alien potato pests occur would be high, whereas that of waste from tropical produce imported to temperate climates could be low.

Waste management

It is both good plant health practice and good commercial practice to minimize the creation of waste as much as possible. This can be done in various ways, depending on the type of operation. For example, contracts for purchase of produce can specify that it should carry not more than specified maximum amounts of waste material. By harvesting crops during suitable weather, using appropriate machinery, and cleaning or trimming produce in the field, much soil and plant waste can be left on the production site or, at least, on the producing farm. This avoids dispersal of much waste and having to dispose of it at a later stage in the marketing chain, when disposal may be more difficult and more costly. Where troublesome pests are known to be present in the crop and returning debris to the field would encourage their spread (for example, with *Verticillium* wilt diseases), the waste material may have to be treated or disposed of in some other way.

The volume of waste material for disposal can also be reduced by re-using it in an appropriate way, if necessary after it has received suitable treatment. When re-use of waste material is being considered, the potential phytosanitary risk of the material should be assessed carefully. Washing water and certain types of plant material can be re-used but will normally require some treatment before this can be done. Waste soil, even if carrying a small amount of pest contamination, can often be used in situations where its phytosanitary risk can be reduced or eliminated, provided transport is not too difficult. For example, it can be used for site filling where it is automatically buried under roads or other major paved or built-up areas, it can be used on road and rail embankments and in other amenity areas well away from commercial arable agriculture, or it can be used for improvement of non-arable land such as woodland, permanent grassland or marsh, where erosion is not likely and any pests it contains are unlikely to find hosts. Careful consideration should be given to the possibility that the land use might change in the future. Disposal of waste without re-use by burial or by incineration is usually costly and may carry risks of pollution, so, wherever possible, a satisfactory re-use should be sought, if necessary after appropriate treatment.

Treatments for liquid waste

In most countries the disposal of large quantities of liquid waste (effluent) must conform to discharge consents from the appropriate authority responsible for the environment, water supply or drainage. Such consent will depend on the volume and nature of the effluent and the amount of suspended solids it contains. Usually consents will not be granted for aqueous effluent with a high content of suspended solids, and this must undergo treatment before discharge to the mains sewer or into a watercourse. A much lower content of solids is usually required for discharge to watercourses. Where such effluent is assessed as having a low phytosanitary risk, it is usually treated by preliminary screening, filtration or discharge into a series of settling lagoons, in which the solids can sediment out of suspension. These processes can render the effluent fit for industrial re-use or acceptable for discharge, but will not necessarily remove all plant pests. Small particles of organic matter carrying fungal or bacterial pests, or propagules of the pests themselves, may not sediment out and can pass through simple filters, although populations may be decreased. Effluent that is discharged to sewers feeding public sewage treatment facilities will normally be subject to biological oxidation through filter beds or ditch systems, which convert dissolved nutrients into biomass, assist the removal of solids, and decrease the populations of suspended microorganisms. However, these processes also may not eliminate all plant pests from high-risk liquid waste.

Where effluent assessed as carrying a high phytosanitary risk must be rendered safe for discharge by elimination of serious plant pests, it will need special treatment. There are several options for this, including UV irradiation, heating, microfiltration or treatment with ozone or an environmentally acceptable, non-persistent, disinfectant such as peracetic acid. Each of these treatments has limitations and must be selected according to the nature of the pest to be eliminated. For example, UV irradiation is effective against some bacteria, but less so against fungi, and is not effective for effluent or water with a high content of suspended solids. Heating is usually effective against all pests, but it must reach and maintain a sufficiently high temperature to kill the pests, it is costly and also is normally applied as a batch treatment, not as a continuous flow. Provided high-risk effluent does not also carry environmentally damaging substances, it may be possible to discharge it to tidal waters or to other outfalls where it will not be returned to agricultural land.

Small amounts of liquid waste, such as may be generated during scientific investigations or testing, may present a serious phytosanitary hazard if the waste contains viable soil- or water-borne pests. For example, testing *Solanum* potatoes for resistance to *Synchytrium endobioticum* or strawberries for infection with *Phytophthora fragariae* may generate waste drainage water carrying viable zoospores and possibly also the long-lived resting spores. Such work should be done within units having self-contained drainage, which is not directly connected to the mains sewerage system but drains into a sealed reservoir. This drainage water can then be treated in batches by boiling for 10 min or with a suitable disinfectant before discharge. This process can be automated.

Treatments for solid waste

Suitable treatments for solid waste will depend on its assessed phytosanitary risk, the nature and volume of material, and the type of pest present. Digestion by microorganisms in anaerobic or aerobic processes in sealed containers may be suitable for low-risk material but, depending on the temperatures reached, will only eliminate the least persistent pests. Anaerobic digestion achieves temperatures of 30–40°C, while aerobic digestion may exceed 50°C.

Low-risk plant material can be composted, and this may produce compost acceptable for returning to agricultural land, horticulture or for amenity purposes. Composting results in the biological degradation of soft plant tissues to yield a friable organic material rich in available nutrients. There are many different methods for composting. Some methods rely mainly on the action of worms, which feed largely or exclusively on decaying organic matter (in Europe mainly *Eisenia fetida* and *E. andrei*) and operate at moderate temperatures. Others depend more on the action of microorganisms and generate considerable heat. Only those methods that operate at the highest possible temperatures are effective in eliminating many plant pests. The most effective systems achieve temperatures of 60–65°C for several days, which is sufficient to kill most plant pests and weed seeds. It is essential that the bulk of material undergoing composting is well mixed and turned during the process, so that all parts of the bulk are exposed for a sufficient period to the temperatures generated. The temperature of the bulk being composted should be monitored near the margins as well as more centrally to confirm the temperatures reached. Compost produced by systems that operate at lower temperatures may still carry pest populations, albeit reduced, and must be used with caution, depending on the source of the material used and its known or suspected pest status.

Waste plant material, growing media and soil carrying a high phytosanitary risk can be treated with heat. This can be applied in various ways but, as with composting, it is essential to ensure that the whole bulk of material is exposed to the heat applied and the process should be monitored to check that it reaches and maintains the required temperature for the appropriate length of time. Where facilities are available, and for limited quantities of solid material, boiling in water is an effective treatment. As a general guide, solids up to 2 cm diameter should be boiled for a minimum of 20 min, and those up to 10 cm diameter for a minimum of 30 min. As described earlier in this chapter, steaming is a common means of eliminating pests from horticultural solid waste. It is a convenient method to use where facilities are available and mobile steaming units can be obtained for treating limited quantities of material. Steaming that results in all parts of the bulk reaching 80°C for at least 1 h is normally effective.

Pests are usually more resistant to heat applied to dry material and need both a higher temperature and a longer treatment period for successful elimination. A minimum temperature of 120°C for at least 1 h should eliminate all plant pests, provided that all parts of the bulk attain this temperature for the full hour.

Disposal of solid waste

Where treatment of solid waste to render it re-usable is impracticable, too costly or would not reduce the phytosanitary risk to acceptable levels, permanent disposal is unavoidable. Dumping at sea is ecologically damaging and must be avoided. There are two main alternatives: deep burial or incineration. Disposal by deep burial should be done only at a landfill site officially approved and appropriately licensed to receive such waste. Among other factors, approval will normally take into account the situation of the site and the security of the material deposited, drainage and risk of contamination of groundwater, and the competence of management. Sites from which seepage might enter groundwater or watercourses, or which might lose contaminated material by rain run-off, wind action, or dispersal by animals, birds or humans are unsuitable. As described in Chapter 7, small amounts of material may be transported and buried in sealed plastic bags, but larger quantities may need to be transported in high-sided trucks, which are sheeted over and sealed. The management responsible for the waste site should be given advance warning concerning the nature of the waste and its contamination so that appropriate action can be taken. On arrival at the disposal site such material should be buried as soon as possible and the landfill should be capped finally with at least 2 m depth of uncontaminated soil (Fig. 12.1). Transport vehicles must be thoroughly cleaned and decontaminated.

Fig. 12.1. Waste-handling machines at work on a landfill waste disposal site. Note attendant birds (which may remove infective debris), and that the land in the foreground has been capped with soil preparatory to restoring it to agricultural or other use. (Photo: courtesy of Environment Agency.)

Incineration is effective in eliminating plant pests, provided that all the contaminated waste material is burnt. Some countries have legal controls on burning of waste, but in any case incineration of contaminated plant material should be properly managed by responsible persons. As incineration can create serious pollution, it is best done in specialized waste incinerators. Special care should be taken to ensure that no contaminated material is carried up into the atmosphere and that any smoke generated presents no hazard.

Where the waste is largely composed of soil or other growing media, volumes are large, and the pest risk is assessed as low, it may be acceptable to dispose of it to non-agricultural sites. For example, it could be used for in-fill on construction sites, for road or rail embankments, for use in urban amenity areas, or for spreading on non-agricultural or non-arable land, such as uncultivated grazing areas or forestry plantations.

Containment facilities

Containment facilities may be required for many plant health purposes. For example, they may be needed to house and prevent the escape of licensed plant pests imported for research or screening trials (Chapter

6) or to hold imported plants of known or unknown health status in post-entry quarantine, preventing the escape of any pests they may harbour so that they can be freed from infections and later used for breeding or propagation. In some situations containment facilities may also be needed to hold free from infection the nuclear stock used for supplying propagation material to certification schemes (Chapter 8). The type of containment facilities required will depend on the nature of the material to be held and the phytosanitary risk, either in possibly disseminating pests or in sustaining incoming infection. For plant pests, containment will normally be in the laboratory, growth chamber or glasshouse, while plants will normally be held in protected environments of some kind. Occasionally it may be safe to hold plant material in the open.

The paragraphs below outline the kinds of facilities and conditions commonly used to contain the types of biological material mentioned above and reduce the phytosanitary risk to acceptable levels. However, many other physical and managerial safeguards could be employed to meet special risks posed by certain material of particular plants or plant pests. GMOs, for example, may require special conditions but are not specially considered here. Practical safeguards, as used by plant health authorities and institutions in many countries, including detailed layout plans for containment facilities, are described by contributors to Khan and Mathur (1999), particularly by Mears and Khan (1999). Traynor *et al.* (2001), the text of which is also available at the website www.isb.vt.edu, provide detailed guidance for the containment of GMOs in greenhouses under USA conditions. However, much of this is also applicable elsewhere and to the containment of organisms presenting a phytosanitary risk.

Containment of plant material

QUARANTINE UNITS. Plant quarantine facilities do not need large areas of land, but should be in areas well isolated from crops related to the imported plant species being held. Town sites can be suitable. They should have secure boundaries with adequate perimeter fences or walls to prevent unauthorized access and usually will need both an area for growing plants in the open, protected accommodation, and laboratory facilities. Quarantine facilities for plants are commonly operated by the NPPO but may be operated privately under licence or by approval. In either case, good management and satisfactory technical expertise are essential. Overall responsibility for the premises should be vested in one senior person, who should maintain liaison with the NPPO and keep detailed records, including dates and times of personnel entry, incoming and outgoing plant material, and treatments applied. Electronic swipe cards can conveniently restrict access to any area and also automatically provide a record of times of entry and exit. Premises should be clean, well organized and efficient, not only for the purposes of minimizing escape of quarantine or alien organisms or cross-contamination of material held, but should be seen to be so to engender confidence in clients and trading partners in both the domestic and foreign commercial spheres and government services. Regular official inspections (for which a checklist of items is helpful) and good maintenance of facilities and equipment are necessary. Adequate office, storage, laboratory and plant-handling space is helpful in maintaining order and efficiency. An emergency generator, soil and plant washing facilities, and access to an incinerator will also usually be necessary. Quarantine premises should display a conspicuous notice of their title. At the entrance and within the facilities there should be further signs restricting entry to designated personnel and designating the use of particular rooms or reminding workers of necessary hygiene precautions. Non-essential visitors to quarantine facilities should be banned or restricted, according to the phytosanitary risk.

Plant quarantine facilities will normally be equipped with greenhouses (glasshouses, 'poly-tunnels' or other protective structures). A modular design is often convenient, although special structures are not always required provided isolation and security are adequate to counter the assessed plant health risk and that no relevant vectors can gain access to the material. For example, many apple virus diseases have no known vectors and, depending on their provenance, imported apple trees might safely be grown in the open during the quarantine period. Separate specialized facilities may be required for handling seeds and growing them on.

Containment structures should be built to a high standard and be more than sufficiently robust to withstand the most extreme weather conditions likely to occur. Severe weather frequently causes damage to commercial glasshouses and tunnels, so structures of equivalent strength would be inadequate. Natural or artificial windbreaks may be advantageous, but the structures should not be near tall trees or dense foliage. An area of about 2 m width around the structure should be kept clear of all vegetation, perhaps by covering with paving, tarmac, or other impervious material. Beyond this a wider area of mown grass helps to keep down local pest populations in the vicinity. Joints between glass, partitions and walls should be sealed and there should be no unsealed gaps between glass or plastic panels and retaining frames. Within the main structure it is convenient to have separate compartments to house different types of material, or for different conditions, and these also need sealed joints and close-fitting doors. Entrance to the main structure should be through an air lock created by a vestibule with both inner and outer well-fitting doors. Ideally, it should not be possible for both inner and outer doors to be open simultaneously. Alternatively, an audible alarm can be fitted to sound if both inner and outer doors are open at the same time. A further refinement, especially when containing mobile insect pests, is to keep the vestibule dark. The main door and doors to compartments should be lockable; combination locks are often convenient. One vestibule may serve several units or compartments and should serve as an exit as well as an entrance, although in large or high-security containment facilities an alternative emergency exit will be necessary. Sometimes it is also good to furnish the vestibule with washing facilities. Foot-operated taps are an option for high-security containment units. Disposable or easily washable protective clothing should be available for use, including disposable covering for (or alternative) footwear. Otherwise a foot pad containing a broad-spectrum disinfectant should be placed at the external door. If necessary, protective clothing can be colour coded for use in different units. There should be bins with plastic bags to receive used disposable clothing and, in facilities dealing with invertebrates, a small deep freeze conveniently serves this purpose, killing or immobilizing any invertebrates that may be contaminating the clothing.

Within glass or plastic-covered structures, plants should not be grown in the existing soil but in containers of some kind, supported on benches if convenient. The floor of the structure should preferably be solid and easily washable, although it may be acceptable to use stout floor coverings in some circumstances. New or cleaned and disinfected growing containers and packaging should be used, with new or sterilized soil-less growing medium, and each container should be clearly labelled. Commercial composts are often suitable but soil, even if partially sterilized, is inadvisable. Even large growing containers should be isolated from local soil and stood within individual drainage saucers or trays in such a way as to avoid splash or contact between neighbouring plants or containers. However, several containers of similar plants can be treated as a lot and retained within a single drainage tray. Watering should normally be from below to avoid cross-contamination by splash. Capillary mats should be replaced after each crop of plants. Automatic watering arrangements will reduce the frequency of visits by maintenance personnel.

Benches should be at a convenient height and easily cleanable. At least some benches should be available that have controlled drainage with arrangements for safe disposal or treatment of drainage water (see p. 242). The water supply should come from a chemically and biologically uncontaminated source, such as a mains supply, a deep borehole, or rainwater retained in a covered tank. Supply from rivers or other surface watercourses should be avoided, especially in agricultural areas. Precautionary treatment of the water supply with UV radiation may be advisable in some circumstances to minimize bacterial contamination (for example, in production of *Solanum* potato planting material free from contamination with *Erwinia carotovora*).

Positive or negative pressure, if required to prevent the entrance or escape of air-borne organisms, can be provided by fans with screens or (for high security) high-efficiency particulate air (HEPA) filters. This can be done on a small scale with bench-top cabinets where appropriate. In warm climates, or when the contained plants require a cold period, it is often difficult to prevent temperatures within protective structures escalating above acceptable levels. In these situations, increased ventilation can be provided by use of fine mesh gauze screening to cover part or all of the structure, fans to increase the throughput of air, or a cooling system using evaporative or mechanical air conditioning may be installed. Controlled shading is also helpful. Gauze screening is easily damaged and must be inspected frequently and repaired promptly where necessary. Depending on colour and mesh size, it may also reduce incoming light levels more than glass or plastic sheet, and even the smallest available mesh sizes usually are not impermeable to spores or very small invertebrates such as thrips. There should be well-fitted screens or brushes at vent openings. Particular attention should be given to the gaps around doors (which may need to be fitted with flanges) and to the thorough cleaning and disinfection of compartments when each project has been completed or after the plants have been removed.

New introductions to quarantine facilities should be held in a separate structure to those already tested and found free from significant pests. For seeds and other robust material, routine fumigation or cold treatment may be a helpful precaution before consignments are opened. Disposable knives or scalpels should be used where appropriate and, depending on circumstances, there should be separate sets of implements for each house, compartment or crop, which are cleaned and disinfected after each batch of operations. Disposable gloves should be worn or hands washed before each operation. Where appropriate, a suitable pest control regime should be applied as a precaution to give effective control of pests that might be expected to occur, but it should not mask the symptoms of pests for which plants are being quarantined. Where possible, plants in quarantine should not be allowed to flower, or should at least be prevented from disseminating pollen. Where tubers or cuttings are stored before planting, they should be kept free from pests (such as aphids or mites) by suitable treatment.

CONTAINMENT OF NUCLEAR STOCK. Certification schemes for the production of healthy planting material from pathogen-tested stock (Chapter 8) are generally mainly concerned with minimizing infections by common pathogens. The pathogens against which the nuclear stock must be protected are therefore mostly non-quarantine pests and the containment conditions usually do not need to be of very high security. The protection required is against incoming infection and not for the prevention of pathogen escape. In some cases, where sites can be found that are suitably isolated and with climates unfavourable for infection by the relevant pathogens, nuclear stock can be maintained in the open. Maintenance in the open will also be possible where the relevant pathogens are not aerially transmitted (viruses transmitted by contact or grafting, for example). Where more secure containment of pathogen-tested nuclear stock is needed, a good-quality glasshouse or gauze (screen) house with some of the features mentioned for quarantine units will usually be sufficient. Air-lock entrances and screens over vents and fans would be needed for protective structures. The application of an effective pesticide regime provides an additional safeguard in all these conditions.

Where nuclear stock of tree fruit plants is maintained within protective structures, attention must be given to the quality and quantity of the bud or graft-wood produced. Insufficient light or cold may result in poor ripening of the wood and poor-quality propagation material.

Containment of plant pests

For the USA, details of facilities and practices for different biosafety levels are available on the National Institutes of Health website, www4.od.nih.gov/ and for greenhouses are described by Traynor *et al.* (2001). The management and many of the safety practices described above for plants will also apply to containment of plant pests. Normally, cultures of high-risk microorganisms will be required to be contained *in vitro* within a lockable incubator in a lockable laboratory with restricted access. Where maintained on plant hosts, pests may be contained in secure growth cabinets or chambers with controlled temperature, light and (where necessary) humidity. Insects and mites are normally contained in cages of fine gauze or ventilated boxes held within such cabinets or chambers. It is advantageous to be able to carry out manipulations without opening the cages. Where sleeves of fine gauze or other material are fitted for this purpose, these must be checked regularly for wear. In some cases containers may need to be surrounded by a cool area to inhibit the mobility of small invertebrates such as thrips or mites. Traps, such as yellow or blue sticky traps or electric UV traps, also should be suitably positioned to guard against escape and to monitor security. Monitoring may also include susceptible trap plants positioned where escape might occur. Lower-risk organisms are usually held in similar conditions but security may be relaxed in proportion to the phytosanitary risk.

Transfers or inoculations with microorganisms should be done in isolated clean rooms reserved for the purpose and, where appropriate, within laminar-flow safety cabinets using normal microbiological sterile techniques. Other precautions such as HEPA filters, air conditioning, negative pressure rooms, air-lock entrances, the use of protective clothing and decontamination showers may be necessary for dealing with high-risk organisms. Pest containment facilities should also have dedicated areas for washing and sterilizing equipment, with autoclaves and ovens for sterilization or disposal of equipment or waste material as necessary. Entomological waste should be frozen in a deep-freeze prior to autoclaving or incineration. Incineration facilities should be available, but not necessarily on site. Floors should be smooth, hard and washable, and controlled drainage may be necessary where certain water- or soil-borne pests are held.

Large-scale soil movement

Major civil engineering projects often necessitate the movement of soil on a large scale. The construction of large or extensive buildings, roads, railways and pipelines are all examples that involve the large-scale movement of soil, and where this is to be done in agricultural areas there may be a risk of spreading soil-borne plant pests. The NPPO will need to check from records held whether any serious soil-borne pests, or any areas on which phytosanitary restrictions have been imposed, occur in the area of operations. If so, a risk analysis should be done to determine whether precautions are necessary to minimize the risk of spreading soil-borne pests. Even where pests are not known or suspected to occur, it is good practice to minimize the movement of soil as much as possible, to replace as much of the soil as possible in its former position, and to avoid the mixture of soil from different places. It is usually possible to remove the topsoil (which will contain virtually all of any pest population) and retain it separately nearby on site. Soil-handling machinery should be well cleaned and, if necessary, disinfected on site before moving to other areas. Where contaminated or suspect soil must be removed from the site, this should be done as described above for disposal of solid waste.

Record keeping by plant health authorities

Record-keeping on a larger and longer scale than that needed for commercial businesses is also an essential part of good practice for plant health authorities. These need to be aware of businesses that deal in plants or other items subject to phytosanitary regulations, their addresses, the names of people in charge, and the type and extent of their operations. Permanent records with accurate maps must also be kept of land scheduled on account of outbreaks of soil-borne pests, and the particular restrictions applying in each case (see Chapter 7). To collate this information it is convenient to operate a registration system for such businesses so that, in the process of registration, all essential information can be obtained and held in a convenient way. In many countries the holding of such information is itself subject to legal requirements, which may govern the type of information that may be held, its security and the purposes for which, or to whom, it may be divulged.

Appendix I: The Regional Plant Protection Organizations

The Regional Plant Protection Organizations (RPPOs) provide coordination at a regional level for the activities and objectives of the International Plant Protection Convention (IPPC, see Chapter 3) and for more local plant protection and plant health activities. In 2002 there were nine RPPOs covering the areas shown in Fig. 3.2. Information on each of these is summarized below. Some RPPOs are much more active than others and the information available varies greatly. Where available, the respective websites may provide more up-to-date information. Links to and information on the RPPOs can also be found on the United Nations Food and Agriculture Organization (FAO) website, www.ippc.int/cds_ippc/IPP/En/default.htm and on the European and Mediterranean Plant Protection Organization (EPPO) website, www.eppo.org/

Asia and Pacific Plant Protection Commission (APPPC)

Region

East Asia, the Indian Subcontinent, Australasia and the Pacific.

Secretariat

c/o FAO Regional Office for Asia and the Pacific, Maliwan Mansion, 39 Phra Atit Road, Bangkok 10200, Thailand.
Tel.: +66-2-281-7844, ext. 268; fax: +66-2-280-0445;
e-mail: chongyao.shen@fao.org

Languages

English.

Member countries

Australia, Bangladesh, Cambodia, China, Fiji, France (French Polynesia), India, Indonesia, Laos, Malaysia, Myanmar, Nepal, New Zealand, Pakistan, Papua New Guinea, Philippines, Republic of Korea, Solomon Islands, Sri Lanka, Thailand, Tonga, Viet Nam, Western Samoa.

Notes

Established 1956. The organization has a chairman, executive secretary and executive committee, and a secretariat.

Caribbean Plant Protection Commission (CPPC)

Commission de la Protection des Plantes dans les Caraïbes; Comision de Proteccion Fitosanitaria para el Caribe.

Region

Caribbean.

Secretariat

c/o FAO Sub-Regional Office for the Caribbean, PO Box 631-C, Bridgetown, Barbados.
Tel.: +246-4267110/1; fax: +246-4276075;
e-mail: gene.pollard@field.fao.org

Languages

English, French, Spanish.

Member countries

Barbados, Colombia, Costa Rica, Cuba, Dominica, Dominican Republic, France (Guadeloupe, French Guiana, Martinique), Granada, Guyana, Haiti, Jamaica, Mexico, Netherlands Antilles, Nicaragua, Panama, Puerto Rico, St Kitts and Nevis, St Lucia, Suriname, Trinidad and Tobago, United Kingdom (British Virgin Islands), USA, Venezuela.

Notes

Established 1967.

Comité Regional de Sanidad Vegetal del Cono Sur (COSAVE)

Region

Southern South America.

Secretariat

Presidencia del COSAVE (Comité Directivo), Millán 4703, CP 12900, Montevideo, Uruguay.
Tel.: +598-2-309-2219; fax: +598-2-309-2074;
e-mail: cosave@mgap.gub.uy; website: www.cosave.org.py/baseesp.htm

Languages

English, Spanish.

Member countries

Argentina, Brazil, Chile, Paraguay, Uruguay.

Notes

COSAVE originated in 1980 as an *ad hoc* committee initiated by the Directors of Plant Protection of the five member states. The present name and constitution were agreed in Montevideo in 1989 and by 1991 had been ratified by all member states. The administrative structure and regulations were approved during 1991 and regular activities started in 1992. The COSAVE Presidency and the Technical Secretary are intended to rotate every 2 years among the member states, but sometimes it is convenient for the Technical Secretary to remain in a different country to the Presidency. Finance is through annual subscriptions from member states and is administered by the Inter-American Institute for Co-operation in Agriculture (IICA). COSAVE is now a recognized RPPO within the framework of the IPPC. Its main objectives are to prevent the introduction and spread of agricultural pests, to minimize their impact on agricultural production, and to harmonize phytosanitary measures in order to facilitate regional and international trade in plants and plant products. Since the World Trade Organization Agreement on the Application of Sanitary and Phytosanitary Measures (WTO-SPS) in 1994, COSAVE has started to develop regional standards to harmonize phytosanitary regulations and procedures. COSAVE formulates standards that are not obligatory but provide general guidelines for the community legislation developed by MERCOSUR for the commercial sector. There are now a considerable number of COSAVE regional standards, which are grouped under the headings of Organization and

Operations, References, Phytosanitary Measures, Biological Control, Certification of Materials for Plant Propagation, Phytosanitary Products, Procedures and Analytical Methods. These, and much other information, are available through the website in English, Portuguese and Spanish.

Comunidad Andina (CA)

Region

North-western South America.

Secretariat

Casilla Postal 18-1177, Lima 18, Peru.
Tel.: +511-411-1400; fax: +511-411-3329;
website: www.comunidadandina.org/

Languages

English, Spanish.

Member countries

Bolivia, Colombia, Ecuador, Peru, Venezuela.

Notes

Established 1969. The Comunidad Andina (Andean Community) is a sub-regional organization of international legal status, principally aiming to promote the economic and social development of its member countries and eventually to create a Latin American common market. Since its inception, it has developed a range of institutions for its organization and administration, including the Andean Presidential Council, the Commission of the Andean Community, the General Secretariat of the Andean Community, Andean Parliament, the Court of Justice of the Andean Community and a Free Trade Zone. An office within the Secretariat of the Andean Community is responsible for international and regional phytosanitary affairs.

European and Mediterranean Plant Protection Organization

Organisation Européenne et Méditerranéenne pour la Protection des Plantes (OEPP).

Region

Europe and Mediterranean.

Secretariat

1 rue le Nôtre, F-75016 Paris, France.
Tel.: +33-1-4520-7794; fax: +33-1-4224-8943;
e-mail: hq@eppo.fr; website: www.eppo.org/

Languages

English, French, Russian.

Member countries

Albania, Algeria, Austria, Belarus, Belgium, Bulgaria, Croatia, Cyprus, Czechia, Denmark, Estonia, Finland, France, Germany, Greece, Guernsey, Hungary, Ireland (Republic of), Israel, Italy, Jersey, Jordan, Kyrgyzstan, Latvia, Lithuania, Luxembourg, Macedonia (Republic of), Malta, Morocco, The Netherlands, Norway, Poland, Portugal, Romania, Russia, Slovakia, Slovenia, Spain, Sweden, Switzerland, Tunisia, Turkey, Ukraine, United Kingdom.

Notes

Established 1951. EPPO is an independent inter-governmental organization and receives no administrative support from FAO. As well as its functions under the IPPC, its main aims are to advise and assist member governments on the technical, administrative and legislative aspects of operating an efficient and effective phytosanitary service, to develop regional standards for phytosanitary measures, to promote the harmonization of phytosanitary regulations in the region it covers, and to promote the use of modern, safe and effective pest control methods according to the principles of good agricultural practice. Its task is somewhat complicated by the fact that its region includes the European Union, which also has strong interests in plant health and quarantine (see Chapter 4). EPPO is governed by a council, comprising representatives of member governments, meeting annually, and an executive committee. There is a secretariat with a staff of about ten responsible to the Director General. EPPO operates mainly through two working parties: the Working Party on Phytosanitary Regulations and the Working Party on Plant Protection Products. Reporting to each of these working parties are a number of technical panels covering particular subject areas. Examples of these are the Panels on Bacterial Diseases, Diagnostics, Phytosanitary Measures, Quarantine Pests for Forestry, and Safe Use of Biological Control, reporting to the Working

Party on Phytosanitary Regulations, and the Panels on Efficacy Evaluation of Fungicides and Insecticides, Environment Risk Assessment, and Good Plant Protection Practice, reporting to the Working Party on Plant Protection Products. Panels are formed or dissolved as necessary to deal with topics as their importance fluctuates, and each working party has about 5–15 panels in operation at any one time. Quarantine pests are reviewed and classified on the A1 List (for those not occurring in the EPPO area) or A2 List (for those present in the EPPO area). For each pest EPPO aims to publish a Data Sheet (Smith *et al.*, 1996), illustration, and a Pest-specific Phytosanitary Requirement, setting out what phytosanitary measures are recommended to prevent its introduction and spread. EPPO convenes scientific meetings on matters of current phytosanitary concern and operates a documentation and information service. It publishes a wide range of material, both as hard copy and electronically, including books, conference proceedings, computer software and the scientific journal *EPPO Bulletin*. A large amount of information is available on-line through the website, including *EPPO News* and the EPPO Reporting Service (containing monthly reports of items and events of plant quarantine concern). Software includes the EPPO Plant Quarantine Data Retrieval System (PQR) and the EPPO Database System on Phytosanitary Regulations (PRS). Besides scientific papers, the *EPPO Bulletin* publishes many other kinds of information, including reports of meetings, datasheets on quarantine organisms, phytosanitary procedures, certification schemes for producing healthy vegetative planting material, guidelines for efficacy evaluation of plant protection products, and pest risk analysis (PRA) guidelines.

Inter-African Phytosanitary Council (IAPSC)

Conseil Phytosanitaire InterAfricain (CPI).

Region

Africa, Madagascar.

Secretariat

PO Box 4170, Nlongkak, Yaoundé, Cameroon.
Tel.: +237-222528; fax: +237-224754.

Languages

English, French.

Member countries

Algeria, Angola, Benin, Botswana, Burkina Faso, Burundi, Cameroon, Cape Verde, Central African Republic, Chad, Comoros, Congo (Democratic Republic of), Congo (Republic of) Côte d'Ivoire, Djibouti, Egypt, Equatorial Guinea, Ethiopia, Gabon, Gambia, Ghana, Guinea, Guinea-Bissau, Kenya, Lesotho, Liberia, Libyan Arab Jamahiriya, Madagascar, Malawi, Mali, Mauritania, Mozambique, Namibia, Niger, Nigeria, Rwanda, São Tomé and Principe, Senegal, Seychelles, Sierra Leone, Somalia, South Africa, Sudan, Swaziland, Tanzania, Togo, Tunisia, Uganda, Zambia, Zimbabwe. (Membership should include all Organization of African Unity (OAU) member states.)

Notes

Established in 1954 by a Convention responding to the FAO call for regional programmes to control post-harvest losses. Originally named the Interafrican Phytosanitary Commission, covering Africa south of the Sahara, its headquarters were located at the Commonwealth Institute of Entomology in London, UK. Since then it has undergone several changes in structure, association and title. It was taken over in January 1960 by the Technical Co-operation Commission in Africa South of the Sahara which, in 1965, became the Scientific, Technical and Research Commission (based in Lagos, Nigeria). At the same time, this body became a part of the OAU and the Commission became part of the OAU Scientific, Technical and Research Commission, extending its cover to all states of the OAU. In 1967, on the recommendation of the OAU Heads of States and Governments, the headquarters of the Commission were transferred to Yaoundé, Cameroon. At the meeting held in September 1967 at Kinshasa, the OAU Council of Ministers adopted the Inter-African Plant Protection Convention (Phytosanitary Convention for Africa) and agreed that it should apply to all member states (Nkouka, 1992). In 1969, the name of the Commission was changed to Inter-African Phytosanitary Council (IAPSC) by the OAU General Secretariat.

The objectives of the IAPSC are designated by the OAU and, besides phytosanitary responsibilities, it aims to stop uncontrolled commercialization and dissemination of agro-pharmaceutical chemicals in Africa, and to prevent damage to human health by exposure to toxic chemicals in pesticide factories and treatment works. It guides member states on technical, administrative and legislative measures against plant pests (including noxious weeds), obtains and disseminates information, encourages cooperation on plant protection and phytosanitary matters, and collaborates with other international organizations with similar aims. The IAPSC is governed and administered by the General Assembly, an executive committee of ten members, an advisory scientific council, and a scientific secretariat headed by the Scientific

Secretary, with about six staff. Responsibilities include phytosanitary training for member states, organization of conferences and scientific meetings, various development projects and publications. The latter include the *African Journal of Plant Protection*, a *Quarterly News Bulletin*, and Co-ordinated Phytosanitary Regulations for Africa. Implementation of the Inter-African Phytosanitary Convention varies among member states of the OAU and has not been uniformly applied.

North American Plant Protection Organization (NAPPO)

Organisation Nord-Américaine pour la Protection des Plantes; Organizacion Norte-Americana de Proteccion a las Plantas.

Region

North America.

Secretariat

Observatory Crescent, Bldg. #3, Central Experimental Farm, Ottawa, Ontario K1A 0C6, Canada.
Tel.: +1-613-759-6132; fax: +1-613-759-6141;
e-mail: imcdonell@em.agr.ca; website: www.nappo.org/

Languages

English, French, Spanish.

Member countries

Canada, Mexico, USA.

Notes

Established 1976. NAPPO coordinates the efforts of Canada, the USA and Mexico to protect their plant resources from the entry, establishment and spread of regulated plant pests while facilitating trade. In doing this, account is taken of the interests of industry and the environment. Staff: Executive Secretary, Administrative Assistant, and consultants as required. Activities are directed by an executive committee, consisting ex-officio of the heads of member countries' national plant protection organizations. In 2002 there were Panels on: Accreditation, Biocontrol, Biotechnology, Citrus, Forestry, Fruit, Fruit trees, Grain, Grapevine, PRA, Pest Alert System, Potato, Seeds, and Standards. The Executive Secretary is responsible to the Executive Committee. He administers day-to-day operations and leads the Working Group. This consists of a representative appointed by each member country. The Working Group is the main body that drives NAPPO activities. It gives advice and makes recommendations to the Executive Committee on policy, procedures and regional phytosanitary standards. It also coordinates activities of the panels, produces position papers on phytosanitary issues of current concern, monitors the implementation of NAPPO policies, procedures and standards, and reports at the NAPPO annual meeting and meetings of RPPOs. An important activity is the development and approval of regional standards for various aspects of phytosanitary work, and a considerable number of these have been approved or are in draft. These regional standards are consistent with the provisions of the IPPC and the WTO-SPS Agreement, and are designed to be more widely adopted.

Organismo Internacional Regional de Sanidad Agropecuaria (OIRSA)

Region

Central America.

Secretariat

Calle Ramon Belloso, Col. Escalon, San Salvador, El Salvador. Tel.: +503-263-1123/4/5; fax: +503-263-1128; e-mail: oirsa@ns1.oirsa.org.sv; website: http://ns1.oirsa.org.sv/

Languages

English, Spanish.

Member countries

Belize, Costa Rica, El Salvador, Guatemala, Honduras, Mexico, Nicaragua, Panama.

Notes

In 1947 the seven states of Costa Rica, El Salvador, Guatemala, Honduras, Mexico, Nicaragua and Panama united their resources and experience to combat a devastating outbreak of locusts, which was threatening the crops of the region. With this objective, the seven ministries of agriculture created the Comité Internacional para el Combate de Langosta. The success of this exercise prompted the creation of a permanent international coordinating body, and OIRSA was established in 1953 with the following objectives:

- to promote the modernization and strengthening of the structures concerned with the protection and health of agriculture in member states;
- to coordinate action for the prevention, control and eradication of pests and diseases of socio-economic importance;
- to develop programmes for training and dissemination of information;
- to advise on the harmonization of laws, regulations and orders concerning agricultural health;
- to support the globalization of free trade;
- to support research on plant and animal health;
- to coordinate with the private sector the identification and solution of health problems;
- to administer the Quarantine Treatment Service for plants, animals, products, by-products, and means of transport.

OIRSA now comes under the Comité Internacional Regional de Sanidad Agropecuaria (CIRSA), which is advised by a technical commission. OIRSA is headed by an executive director, to whom the directorates of plant health, animal health, regional coordination, and administration and finance report. The Executive Director is appointed by CIRSA, which is composed of the ministers of agriculture and livestock of the member states, and formulates policy with the advice of the Technical Commission, which is composed of the directors of the animal and plant health services of the ministries of agriculture of the member states. There is a representative in each member state who coordinates technical and administrative activities. Unlike other RPPOs, OIRSA conducts operational activities in both plant and animal health. For this the technical, administrative and support staff at headquarters and in the member states total about 300 persons.

Within OIRSA the Regional Co-ordination of Quarantine Services has been created to advise, coordinate and support the national quarantine services of the ministries of agriculture and livestock of the member states with the aim of preventing the entry and establishment of new pests in the region. The Regional Co-ordination of Quarantine Services includes the International Quarantine Treatment Service (SITC), Agrolivestock Technical and Quarantine Advice, quarantine stations and other activities.

Pacific Plant Protection Organization (PPPO)

Organisation Phytosanitaire pour le Pacifique.

Region

Australasia, Pacific.

Secretariat

Plant Protection Service Secreatriat of the Pacific Community, Private Mail Bag, Suva, Fiji.
Tel.: +679-370733; fax: +679-370021;
e-mail: mickL@spc.org.fj; website: www.spc.org.nc/

Languages

English, French.

Member countries

Australia (including Norfolk Island), Cook Islands, Fiji, France (for Wallis and Futuna Islands), French Polynesia, Guam, Kiribati, Marshall Islands, Micronesia (Federated States of), Nauru, New Caledonia, New Zealand, Niue, Northern Mariana Islands, Palau, Papua New Guinea, Western Samoa, Solomon Islands, Tokelau, Tonga, Tuvalu, UK (for Pitcairn), USA (for American Samoa), Vanuatu, Wallis and Futuna Islands.

Notes

Established in 1995 under the auspices of the Secretariat of the Pacific Community to coordinate phytosanitary matters. It provides a service including technical advice, training and assistance in all aspects of plant pest control and plant quarantine, from which member countries and territories can draw as they wish.

Appendix II: International Convention on *Phylloxera*, 1878

Translation of the French text reproduced in *Actes de la Conférence Internationale de Phytopathologie, 24 Février–4 Mars 1914.* Institut International d'Agriculture, Roma, 1914, pp. 237–241.

Convention Concerning the Measures to be Taken Against *Phylloxera vastatrix* Signed at Berne, 17th September 1878

His Majesty the Emperor of Germany, the King of Prussia, His Majesty the Emperor of Austria, the Apostolic King of Hungary, His Catholic Majesty the King of Spain, the President of the Republic of France, His Majesty the King of Italy, His most faithful Majesty the King of Portugal, the Swiss Confederation, considering the increasing destruction by Phylloxera and recognising the opportunity for common action in Europe for stopping, if possible, the progress of the malady in the affected countries, and for trying to preserve these regions until the day is saved, after having taken note of the Acts of the International Phylloxera Congress which met in Lausanne from 6 to 18 August 1877, are resolved to conclude a Convention with this aim, and have nominated as their representatives the following:

His Majesty The Emperor Of Germany, King Of Prussia:

- Le Sieur Henri de Roeder, Lieutenant General, Envoy Extraordinary of the Ministry and Plenipotentiary to the Swiss Confederation.

- Le Sieur Adolphe Weymann, Confidential Councillor of the Regency, and Reporting Councillor at the Chancellery of the Empire.

His Majesty The Emperor Of Austria, Apostolic King Of Hungary

- Le Sieur Maurice Baron d'Ottenfes-Gschwind, Envoy Extraordinary and Plenipotentiary Minister to the Swiss Confederation.

His Catholic Majesty The King Of Spain

- Le Sieur Don Narciso de Loygorri, Vicomte de la Véga, Chargé d'Affaires to the Swiss Confederation.
- Le Sieur Don Mariano de la Paz Graëlls, Councillor for Agriculture, Industry and Commerce at the Ministry of [Fomento], Professor of Comparative Anatomy and Physiology at the Central University.

The President Of The French Republic

- Le Sieur Bernard Comte d'Harcourt, French Ambassador to the Swiss Confederation.
- Le Sieur Georges Halua du Frétay, Inspector General of Agriculture.

His Majesty The King Of Italy

- Le Sieur Louis Amédée Melegari, Senator, Minister of State and Envoy Extraordinary and Minister Plenipotentiary to the Swiss Confederation.
- Le Sieur Adolphe Targioni Tozzetti, Professor of Zoology and Comparative Anatomy at the Royal Institute of Higher Practical and Improving Studies of Florence.

His Most Faithful Majesty The King Of Portugal

- Le Sieur João Ignacio Ferreira Lapa, Councillor, Director and Professor, General Agricultural Institute, Lisbon, and Technical Commissioner at the Paris Exhibition in 1878.

The Swiss Confederation

- Le Sieur Numa Droz, Federal Councillor, Head of the Federal Department of the Interior.

- Le Sieur Victor Fation, Doctor of Philosophy (Natural Sciences).

Who after having announced their powers in a proper manner, agreed on the following articles:

Art. 1. The contracting States undertake to complete, if they have not already done so, their national legislation with a view to assuring unified and effectual action against the introduction and the spread of Phylloxera.

This legislation should in particular provide for:

1.1 The inspection of vineyards, gardens, glasshouses and nurseries, the investigations and verifications necessary for research on Phylloxera and operations with the aim of destroying it as far as possible.
1.2 The delimitation of the areas infested by the malady, as the infestation gradually enters or spreads within these States.
1.3 The regulation of transport of vine plants, debris and products of such plants as well as the plants themselves, bushes and horticultural products, in order to prevent the transmission of the infestation out of the infected areas within the State and transmission to other States.
1.4 The packaging and transport of these items, as well as the precautions and arrangements to be taken in cases of infringement of the legislative measures.

Art. 2. Wine, table grapes without leaves or shoots, grape pips, cut flowers, market-garden produce, seeds of all kinds and fruits are allowed to circulate freely in international trade.

Plants, bushes and the various products of nurseries, gardens, glasshouses and orangeries should not be introduced from one State into another except via the customs bureaux designated for that purpose by the adjacent contracting States and under the conditions defined in Article 3.

Uprooted vines and dried vine shoots are excluded from international trade.

Adjacent States agree to the admission into their frontier zones of harvested wine grapes, grape residues, composts, soils, used stakes and vine trainers, provided the said objects do not harbour phylloxera.

Vine plants, cuttings and shoots should not be introduced into a State except with its consent and should not be traded internationally except via the customs bureaux designated and under the conditions of packaging indicated below.

Art. 3. The objects listed at paragraphs 2 and 5 of the preceding article, having been accepted for international trade by the designated customs bureaux, should be accompanied by an assurance under the authority of the country of origin, showing:

(a) that they originate in an area well known not to have been invaded by phylloxera and appearing as such on the special map established and kept up to date in each contracting State;
(b) that they have not been recently imported.

Vine plants, cuttings and shoots should not be traded except in wooden containers completely closed by means of screws, but nevertheless easy to examine and re-close.

Plants shrubs, and the various produce of nurseries, gardens, glasshouses and orangeries must be completely wrapped: stems must be completely freed from soil; they should be surrounded by moss and must, in all cases, be wrapped in wrapping-cloth in such a way as not to allow the escape of any debris and to permit necessary checks.

The Customs Bureau, at its discretion, shall make an examination of the items by official experts, who shall make an official report when they discover the presence of phylloxera.

The said official report shall be sent to the State which is the country of origin in order that the contravention may be followed-up, if there is provision, through the appropriate channels in conformity with the legislation of the said State.

No consignment in international trade should on any account contain vine leaves.

Art. 4. The items stopped by a Customs Bureau for reasons such as not conforming to the packaging conditions prescribed in the preceding article, shall be returned to their point of dispatch at the cost of the owner.

The items on which the experts have detected the presence of Phylloxera shall be destroyed as soon as possible and on the spot by fire, together with their packaging. The vehicles which have transported them shall be immediately disinfected by adequate washing with carbon disulphide, or by any other procedure scientifically recognised to be effective and which shall be adopted by the State. Each State shall take measures to ensure this disinfection is thoroughly carried out.

Art. 5. The contracting States, in order to facilitate co-ordinated action, undertake regularly to communicate to each other:

5.1 the laws and regulations promulgated on the subject by each of them;
5.2 the principal measures taken for implementation of the said laws and regulations, and of the present Convention;
5.3 the reports or extracts from the reports of the various services organised to operate internally and at the frontiers against Phylloxera;
5.4 all discoveries of Phylloxera outbreaks in areas believed to be uninfested and, if possible, of the causes of the outbreak (this communication shall always be made without any delay);
5.5 all maps made for delimitation of uninfested areas and of infested or suspect areas;
5.6 information on the spread of the plague in the areas where it has been detected;
5.7 the result of scientific studies and practical experiences in vineyards with Phylloxera;
5.8 all other documents concerning viticulture from this special point of view.

These various communications shall be used by each of the contracting States for those publications which it shall make on the subject, which publications shall also be exchanged between them.

Art. 6. When it is judged necessary, the contracting States shall be represented at an international meeting to examine the questions which arise from implementation of the Convention and to propose modifications in the light of experience and the progress of science.

The said international meeting shall be held at Berne.

Art. 7. Ratifications shall be exchanged at Berne within the period of six months from the date of signature of the present Convention, or earlier if it is possible to do so.

The present Convention shall enter into force 15 days after the exchange of ratifications.

Any State may comply or withdraw from it at any time by means of a declaration to the Federal High Council of Swizerland, which accepts the mission of serving as intermediary between the contracting States for implementation of Articles 6 and 7 herein.

In testimony of which, the respective Plenipotentiaries have signed it and have appended their official seals.

Done at Berne, the seventeenth day of September of the year one thousand eight hundred and seventy eight.

V. Roeder	G. Halua du Frétay
Weimann	Melegari
Ottenfels	Ad. Targioni
Vicomte de la Véga	Goão Ignacio Ferreira Lapa
Mariano de la Plaz Graëlls	Droz
B. d'Arcourt	Victor Fatio

Appendix III: International Controls on the Use of Plant Pests as Offensive Agents

Introduction

Although the sciences of plant pathology, plant entomology and plant health are largely aimed at controlling the damage done by plant pests to crops, amenity plants and the environment, this knowledge can also be used to employ plant pests for offensive purposes. Some development of weapons based on plant pathogens (and corresponding counter-measures) has been done as a derivative of biological weapons based on human or animal pathogens, and in parallel with the development of chemical weapons, with some of which they have certain features in common, for example, in the mode of delivery.

Plant Pests as Biological Weapons

Before the advent of effective and economic fungicides for cereal crops in the 1970s, it was conceivable that pathogens of cereals causing diseases such as rusts, smuts and blights could be used for strategic attack on a nation's food supply (Van der Plank, 1963). Indeed, three fungus pathogens (*Puccinia graminis*, black stem rust of cereals and grasses; *Puccinia striiformis*, syn. *Puccinia glumarum*, yellow rust of wheat, barley, rye, triticale and grasses; *Pyricularia oryzae*, rice blast) have been declared by the USA as having been developed as part of an offensive capability that was unilaterally abandoned in 1969 (MacKenzie *et al.*, 1985). Since then the number and efficacy of pesticides for major food crops has steadily increased and, in many agricultural systems, their use has become routine, while ever more varieties of the crops themselves have been bred with increasingly effective resistance to more and more species and strains of pest. Efficient crop disease monitoring techniques have been developed and pesticide application mechanisms have also improved enormously.

In addition to these considerations, the difficulties of artificially generating a plant disease epidemic (epiphytotic) are considerable. Plant pests often have very specific requirements for successful attack, with narrow tolerances for temperature, humidity and the stage of host development. They may also require the presence of vectors for transmission, or the presence of a secondary host species to complete the life cycle. The probability of all the necessary conditions being favourable over a wide target area at a particular time is therefore small, even if routine pesticide application has not minimized the risk. Also, the development of an epiphytotic from limited foci could probably be detected and arrested, at least in the more developed countries and provided they were alert. The possibility of international replenishment of food supplies is another factor diminishing the likelihood of a successful strategic attack on food crops.

The vulnerability of the major food crops to artificially generated epiphytotics using wild-type organisms therefore appears to have decreased substantially. However, the possibility of using recent advances in genetic manipulation to enhance the potential of plant pests for serious damage makes it essential to control genetically modified plant pests very tightly. Indeed, the Convention on Biological Diversity (Chapter 3) recognizes that risks to the environment would necessitate this, even with modifications intended for benign purposes.

Economic Damage

Although the danger of overt strategic use of plant pests against food production may have diminished, their possible use on a smaller scale by disaffected groups in covert attacks aimed at economic damage has become a greater risk (MacKenzie *et al.*, 1985). The ease and cheapness with which some plant pests can be manipulated, produced and concealed, and the fact that small quantities could be carried and spread by an individual courier, renders them very suitable for clandestine use. Many countries are economically still heavily dependent on plant products and in these countries serious economic damage could be inflicted by the introduction of certain plant pests on quite a small scale. Those countries whose economies are heavily dependent on perennial crops are particularly vulnerable. Tree crops in particular cannot be quickly replaced, nor can new resistant cultivars be bred rapidly. They usually take several years to reach full cropping capacity and represent a greater investment than annual crops. Crops that are genetically highly homozygous over wide areas are also at greater risk.

It is not always necessary for a plant pest to cause serious physical damage in order to produce severe economic loss. The rise in quality standards for plant produce in domestic and international trade may result in quite superficial damage rendering plant produce unsaleable. In addition, the recognized presence within a country of a serious plant pest may make it impossible to export the produce of the host crop for phytosanitary reasons.

Although serious plant pests will often be subject to control by national and international phytosanitary regulations as quarantine organisms, this may not be so in countries or regions where they are common and widespread, or where the absence of hosts or an unfavourable climate renders them of little concern. Also, the nature of phytosanitary controls for such pests may not be sufficient to prevent their misuse. Other controls may therefore be necessary and international controls on naturally occurring and genetically modified plant pests that could be used for offensive purposes have been agreed.

The Biological Weapons Convention

Although the development of chemical and biological weapons began before the First World War, active development of plant pests as weapons was begun only after the Second World War. This accelerated during the period of the so-called 'cold war' and other nations besides the major powers also started to become involved. Negotiations under the auspices of the United Nations to control these activities were therefore started as part of the major negotiations on general disarmament and control of weapons of mass destruction, and in further extension and development of the Protocol for the Prohibition of the Use in War of Asphyxiating, Poisonous or Other Gases, and of Bacteriological Methods of Warfare, signed at Geneva on 17 June 1925. The latter agreement, often referred to as the Geneva Protocol, banned the use of chemical and biological weapons, but a number of signatories entered reservations, which, in effect, allowed retaliatory use. The Protocol therefore had no effect on the development or production of biological weapons. However, by the late 1960s increasing international opinion that biological weapons should be banned altogether gave impetus to negotiations. These culminated in the Convention on the Prohibition of the Development, Production and Stockpiling of Bacteriological (Biological) and Toxin Weapons and on their Destruction, signed at London, Moscow and Washington on 10 April 1972, commonly known as the Biological Weapons Convention (BWC). The Soviet Union, UK and USA were specified as Depositary State Parties. Further information can be found on website www.acronym.org.uk/bwc/

States that are party to the BWC undertake never, in any

circumstances, to develop, produce, stockpile or otherwise acquire or retain: (i) microbial or other biological agents or toxins that have no justification for peaceful purposes; and (ii) means of delivery for use with such agents or toxins for hostile purposes. Amongst other obligations under the 15 articles of the Convention, the contracting parties also undertake to destroy such weapons and means of delivery, not to transfer such weapons to others or assist their acquisition, and to prevent the development, production or retention of such weapons within territory under their jurisdiction. Article XIV.3 specifies that the BWC would enter into force after ratification by 22 governments, and this took place on 26 March 1975. Under Article XII a review conference was held in 1980, and further review conferences have since been held at approximately 5-year intervals, interspersed with related and preparatory meetings.

One of the shortcomings of the BWC is its lack of an effective verification mechanism. In 1994 the East–West détente after the end of the 'cold war' permitted potential verification measures to be contemplated and examined. By 2001 a draft protocol had been prepared, which included mandatory declarations of certain facilities, visits by an international inspectorate, and provision for States Parties to raise challenges on compliance that could lead to inspection of facilities or areas alleged to be in violation of the Convention. This has not been agreed. However, confidence-building measures have been agreed, which include declarations designed to alert the international community to the occurrence of any significant outbreaks of diseases. These include annual reports by each State Party on the plant disease situation on major crops.

The Australia Group

In 1984 the use of chemical weapons in the Iran–Iraq war and the evident possibility of obtaining precursors for these through the commercial chemical industry raised serious concerns. However, the measures taken by concerned countries to prevent this were far from uniform and this was being exploited in attempts to circumvent the controls. Concerned countries met in Brussels in 1985 to improve cooperation and harmonization of measures, particularly the control of exports to certain destinations. This group of countries (consisting initially mainly of the Organization for Economic Cooperation and Development members, see Chapter 9) subsequently became known as the Australia Group, and further meetings have been held annually in Paris. The Australia Group remains an informal grouping of countries, which has no legally binding obligations, and in the year 2002 there were 34 participants, including the European Commission. Its effectiveness rests solely on the commitment of its participants to prevent the proliferation of chemical and biological weapons by taking effective national measures that also do not significantly impede normal trade in commodities and equipment used for legitimate purposes. Under Article V of the BWC, the parties to the Convention undertake mutually to consult and cooperate in solving any problems that arise in relation to the objectives or the operation of the Convention. All Australia Group participants are parties to the BWC and take a vigorous part in promoting the implementation of its key obligations and confidence-building measures. Further information can be found on the website www.australiagroup.net

Export controls on plant pests

Under Article III of the BWC, States Parties undertake to prevent the transfer of materials that could assist in the manufacture or acquisition of biological weapons. In 1990 increasing evidence that certain dual-use materials and equipment were being diverted for the production of chemical and biological weapons prompted the Australia Group to promote harmonized controls on these items through licensing restrictions on their export. This export licensing system is implemented through national legislation by participating countries and endeavours to ensure that these items are not inadvertently exported to destinations where they could contribute to the proliferation of chemical or biological weapons.

Among the items covered by the Australia Group export licensing system are organisms or toxins that could be potential biological or chemical warfare agents, and chemicals, materials or equipment that could be used for biological or chemical weapons production. Common Control Lists of such chemicals, equipment and organisms have been agreed for this purpose. In addition, for plant pathogens, there is a second category of organisms that are not felt to warrant export control by licensing but demand for which could raise concerns. These are included under Awareness-raising Guidelines.

Criteria

The criteria for placing a plant pest on the lists for export control or Awareness-raising Guidelines have generated considerable debate. It was felt that plant pests that could conceivably be used as biological weapons agents should include those species that: (i) could have severe socio-economic or health impacts, either directly or indirectly; (ii) are easily disseminated; (iii) have a short incubation period; (iv) are not highly host-specific; (v) are easily produced; (vi) have high infectivity

Table III.1. Australia Group Common Control Lists for plant pests.

	Plant pest	Hosts	Diseases caused
Core List for export controls			
Bacteria			
PB1	*Xanthomonas albilineans*	Sugar cane	Leaf scald
PB2	*Xanthomonas campestris* pv. *Citri*	Citrus	Citrus canker
Fungi			
PF1	*Colletotricum kahawae* (syn. *C. coffeanum* var. *virulans*)	Coffee	Coffee berry disease
PF2	*Cochliobolus miyabeanus* (Anamorph *Drechslera oryzae*, syn. *Helminthosporium oryzae*)	Rice	Brown spot and seedling blight
PF3	*Microcyclus ulei* (syn. *Dothidella ulei*)	Para rubber (*Hevea* spp.)	South American leaf blight
PF4	*Puccinia graminis* (forms attacking wheat)	Wheat, other cereals and grasses	Black stem rust
PF5	*Puccinia striiformis* (syn. *P. glumarum*)	Cereals and grasses	Yellow rust, stripe rust
PF6	*Magnaporthe grisea* (Anamorphs *Pyricularia grisea* and *P. oryzae*)	Rice	Blast disease
Genetic elements and genetically modified organisms (GMOs)			
PG1 and PG2	Genetic elements (PG1) or GMOs (PG2) that contain nucleic acid sequences associated with the pathogenicity of organisms in the Core List		
List for Awareness-raising Guidelines			
Bacteria			
PWB1	*Xanthomonas campestris* pv. *oryzae*	Rice	Bacterial blight, kresek disease
PWB2	*Xylella fastidiosa*	Grapevine, peach	Pierce's disease of grapevine, phoney disease of peach
Fungi			
PWF1	*Deuterophoma tracheiphila* (syn. *Phoma tracheiphila*)	Citrus, especially lemon	Mal secco
PWF2	*Monilia rorei* (syn. *Moniliophthora rorei*)	Cacao	Watery pod rot
Viruses			
PWV1	Banana bunchy top virus	Banana	Bunchy top
Genetic elements and GMOs			
PWG1 and PWG2	Genetic elements (PWG1) or GMOs (PWG2) that contain nucleic acid sequences associated with the pathogenicity of organisms in the Awareness-raising Guidelines		

(a low infection dose); (vii) are stable in the environment; and (viii) have no easy, effective treatment for prophylaxis or eradication. It was accepted that while candidate organisms for control should possess many of these characteristics, some might possess only few and yet be suitable for inclusion on one or other of the lists. In particular, it was felt that those that have been the subject of weapons development in the past should be included.

It is desirable that the number of organisms included on the lists is minimized as much as possible, so that tiresome and onerous controls are not applied unnecessarily to numerous organisms or to those frequently used in normal biological research work. Also, there is little point in attempting to control the supply of organisms that are widespread in the wild, even if they could be very damaging. This would omit most common pests of major crops except, possibly, for certain strains that are of limited distribution. However, in contrast, the case for control of genetically modified pathogens is strong. Genetic modification could alter an almost infinite range of characteristics to make an organism more dangerous. Such modifications could include those intended to increase pathogenicity or host range, remove or add the need for a vector in transmission, or could confer the ability to incite the production of toxins. It has been agreed to cover the plant pests shown in Table III.1 by Export Controls or Awareness-raising Guidelines.

It is apparent that the activities of culture collections which trade internationally in cultures of microorganisms could easily fall within the export controls. Such collections in particular, and others trading or exchanging microorganisms internationally, will need to be aware of these controls and how they are applied by the governments of the countries in which they operate.

Use of Plant Pests for Control of Narcotic Crops

For many years microorganisms and invertebrates have been used for the biological control of noxious weeds and plant pests. There are many examples where this has been wholly or substantially successful and the use of biological control in plant health is discussed in Chapter 7. While the BWC prohibits the development or use of biological agents as weapons for hostile purposes, it does not prohibit biological control for peaceful, beneficial purposes. However, it does not contain definitions of 'peaceful' or 'hostile'; neither does it contain any exemption for the use of biological agents for law enforcement purposes. It is also sometimes difficult to distinguish law enforcement from internal civil conflict, and the boundary between 'peaceful' and 'hostile' use may be blurred or subjective.

In recent years the proposed use of plant pests as agents for the control of narcotic crops has raised serious international concerns. These concerns include possible harmful effects on human health and the natural vegetation, and also that the exercise would undermine the BWC, as the production and possibly also the delivery of these agents on a substantial scale could not be distinguished from the production and delivery systems for biological weapons. The facilities and technology used could also easily be switched from benign to hostile use.

Information

Up-to-date information on the current situation on most aspects of biological weapons and their control may be found on the website www.sunshine-project.org The Sunshine Project is an international non-profit-making organization based in Hamburg, Germany, and in Austin, Texas. Its declared aims are to support international treaties and their effective prevention of the development and use of biological weapons and to strengthen global consensus against these, particularly by researching and publishing information and thus raising awareness on a global scale.

Glossary

Definitions marked * follow ISPM No. 5, *Glossary of Phytosanitary Terms* (Anon., 2002a).

allozyme Different forms of an enzyme specified by different alleles at the same gene locus.
APPPC Asia and Pacific Plant Protection Commission
BIP Border inspection post
bolter Tall, coarse, atypical variation of *Solanum* potato plant; root crop plant that is producing a premature inflorescence.
bonsai Dwarf tree or shrub of hardy species grown in shallow container by Japanese method (cf. **penjing**).
British Isles Great Britain together with the island of Ireland and the smaller islands nearby.
CA Comunidad Andina
CBD Convention of biological diversity
certification scheme Quality control system for producing authenticated planting material by means of rules and check inspections, with certificates issued for successful crops (in EPPO terminology including a pedigree requirement for pathogen-tested starting material).
chimera Organ/organism composed of genetically different tissues.
CILSS Comité Permanent Inter-Etats de Lutte contre la Sécheresse dans le Sahel
clone Series of plants derived vegetatively from a common source and genetically identical.
CoAg FAO Committee on Agriculture
commodity A type of plant, plant product or other article being moved for trade or other purpose.*
controlled area A regulated area that an NPPO has determined to be the minimum area necessary to prevent spread of a pest from a quarantine area.*

COREPER Committee of Permanent Representatives (EU)
COSAVE Comité Regional de Sanidad Vegetal para el Cono Sur
CPM Commission on Phytosanitary Measures
CPPC Caribbean Plant Protection Commission
CPVR Community Plant Variety Rights (EU)
DAS-ELISA Double-antibody sandwich-ELISA
Defra Department of Environment, Food and Rural Affairs
derogation An authorized exception to certain EU legislation.
DG Directorate General: largest administrative unit of EU Commission covering a particular policy sector.
diploid Possessing two sets of chromosomes (the usual condition).
Directive Type of EU legislation required to be implemented in all Member States via national legislation.
DLCO-EA Desert Locust Control Organization for Eastern Africa
dodder Parasitic climbing plants in genera *Cuscuta* or *Cassytha*, which are used for transmission of viruses and phytoplasmas.
dunnage Wood packaging material used to secure or support a commodity but which does not remain associated with the commodity.*
DUS Distinct, Uniform and Stable (EU CPVR)
EC European Community
ECE United Nations Economic Commission for Europe
EEC European Economic Community
endemic Native and limited to the area concerned, not being native elsewhere.
EPPO European and Mediterranean Plant Protection Organization (see also OEPP)
equivalence Recognition by EU of measures taken by a third country as being equally effective as those of the EU.
established Not native to the area concerned, but now occurring and surviving in it.
EU European Union
European Commission Executive body of the EU.
European Council The EU Council of Ministers when delegates are Heads of Government.
EUROPHYT EU program for reporting of plant pests.
FAO Food and Agriculture Organization of the United Nations
filiation Pedigree (of material in a certification scheme).
GATS General Agreement on Trade in Services
GATT General Agreement on Tariffs and Trade
GIS Geographical information systems
GM Genetic modification
GMO Genetically modified organism (cf. *LMO*)
Great Britain England, Scotland and Wales.
greenhouse Protective structure for growing plants, including glasshouses, plastic-covered frames, and other structures.
grey literature The agricultural, horticultural and forestry trade and popular press (of trading-partner countries).
HEPA High efficiency particulate air (filters)
heteroecious Describing rust fungi that pass different phases of their life cycle on different species of host plants.

homozygous Having identical alleles at particular loci on each of a homologous pair of chromosomes.
IAPSC Inter-African Phytosanitary Council
ICPM Interim Commission on Phytosanitary Measures
IIA International Institute of Agriculture
IICA Inter-American Institute for Co-operation in Agriculture
import licence Official document allowing import of normally prohibited item and specifying conditions for this.
import permit Official document authorizing importation of a commodity in accordance with specified phytosanitary requirements.*
indexing Routine testing or assessment of plant material for infection or infestation with pests.
indigenous Native to the area concerned.
infected Applies to plants or other living organisms harbouring a parasitic pest (plant pathology). See also **infested**.
infested, infestation Presence in a commodity of a living pest of the plant or plant product concerned. In phytosanitary terminology, infestation includes **infection**.*
interception Detection of a pest during inspection or testing of an imported consignment.*
intranational Within a country (adjective).
IPPC International Plant Protection Convention
isoenzyme Electrophoretically separable forms of an enzyme with identical activities, usually specified by different genes.
ISPM International Standard for Phytosanitary Measures.
ISTA International seed testing association
ITCZ Inter-tropical wind convergence zone
LMO Living modified organism resulting from biotechnology (Convention on Biological Diversity). More generally known as **GMO**.
marketing scheme Quality assurance scheme for planting material, normally having no pedigree requirement.
meristem Tissue consisting of actively dividing cells at the growing points of a plant.
micropropagation Form of tissue culture resulting in new plants, usually using excised shoot tips as starting material.
mitochondrion Subcellular DNA-containing organelle within the cytoplasm where aerobic respiration and energy production occurs.
myco-pesticide Pesticide based on a formulation of fungus spores or propagules pathogenic to the target pest.
NAPPO North American Plant Protection Organization
Notice Official phytosanitary document requiring addressee to take action under specified legislation, or imparting specific information (in UK).
NPPO National plant protection organization,* providing phytosanitary services.
nuclear stock Collection of true-to-variety and healthy plants from which further generations may be propagated.
OECD Organization for Economic Co-operation and Development
OEPP Organisation Européenne et Méditerranéenne pour la Protection des Plantes (EPPO)

off-type Plant showing physical differences from normal true-to-variety type due to mutation.
OIRSA Organismo Internacional Regional de Sanidad Agropecuaria
OSTS Official seed testing station
penjing Non-hardy dwarf tree or shrub grown in container by Chinese method and intended for indoor use (cf. **bonsai**).
Permanent Representation Member State's official delegation to the EU.
pest outbreak An isolated pest population, recently detected and expected to survive for the immediate future.*
Phare Originally the EU programme on Poland/Hungary Aid for the Reconstruction of the Economy, but now covering other European countries.
phenology Study of the ways in which natural phenomena are affected by climate and other environmental factors.
phyto International Phytosanitary Certificate (colloquial).
phytoplasmas Formerly known as mycoplasma-like organisms (MLOs), these are very small bacteria-like pathogens occurring in the phloem tissues and having a variable shape due to lack of a firm cell wall.
plant passport Official document (sometimes in the form of a label) signifying that the item covered conforms to EU phytosanitary regulations and can be moved within the EU, showing its eligibility to move within Protected Zones, and giving information on its provenance.
polyphagous (Of pest) not restricted to one or a few specific hosts.
POSEIDOM Programme of Options Specific to the Remote and Insular Nature of the French Overseas Départments, which are part of the EU.
POSEIMA Programme of Options Specific to the Remote and Insular Nature of the Portuguese territories Madeira and the Azores, which are part of the EU.
PPPO Pacific Plant Protection Organization
PRA Pest risk analysis
Protected Zone Area within the EU in which specified extra measures against certain pests must be taken (denoted ZP).
quarantine Official confinement of regulated articles (including plants) for observation and research or for further inspection, testing and/or treatment.*
quarantine pest A pest of potential economic importance to the area endangered thereby and not yet present there, or present but not widely distributed and being officially controlled.*
RFLP Restriction fragment length polymorphism
ribosome Subcellular RNA-containing organelle, which is the site of protein synthesis.
RNQP Regulated non-quarantine pest: a non-quarantine pest whose presence in plants for planting affects the intended use of those plants with an economically unacceptable impact and which is therefore regulated within the territory of the importing contracting party [of IPPC].*
rogue Plant of different species or variety from that of the main crop.
roguing Removing **rogues**.
RPPO Regional Plant Protection Organization
RT-PCR Reverse transcription PCR

SEM Scanning electron microscope
solarization Solar heating of moist soil by covering with plastic sheeting.
somatic mutation Mutation occurring during division (mitosis) of a non-reproductive cell.
SOP Standard Operating Procedure: detailed, standardized, written instructions and descriptions of procedures to be performed in scientific laboratories.
SPS The WTO Agreement on the Application of Sanitary and Phytosanitary Measures.
stock The plants and propagation material retained for propagating successive crops; also as short for 'rootstock'.
SWARMS Schistocera WARning Management System
taxon A named grouping in the classification of organisms.
TEM Transmission electron microscope
tetraploid Possessing four sets of chromosomes.
third countries Countries that are not members of the EU.
tissue culture Plant propagation by means of growing excised tissue *in vitro* on artificial media.
tolerance The degree to which a particular fault is accepted without resulting in the down-grading or rejection of the material entered (in relation to a certification or marketing scheme).
transparency Showing the reasons underlying the action and without hidden conditions; openness, especially regarding public access to information.
TRIPS Trade-Related Aspects of Intellectual Property Rights
ULV Ultra low volume
UNEP United Nations Environment Programme
United Kingdom Great Britain and Northern Ireland.
UPOV International Union for the Protection of New Varieties of Plants
Vademecum Set of guidelines for plant health operations produced by the EU Standing Committee on Plant Health.
VCU Value for Cultivation and Use (EU CPVR)
VD Variety Denomination (EU CPVR)
viroid Smallest known self-replicating plant pathogen, consisting of naked, usually circular, single-stranded RNA of relatively small molecular weight.
ware Potatoes (or other root vegetable) not intended for propagation.
wilding Atypical *Solanum* potato plant with many weak stems and small leaves with large, heart-shaped terminal leaflets.
WTO World Trade Organization
ZP Zona Protecta; Protected Zone (within the EU).

References and Websites

References

Ainsworth, G.C. (1981) *Introduction to the History of Plant Pathology.* Cambridge University Press, Cambridge.

Anon. (1914) *Actes de la Conférence Internationale de Phytopathologie.* Institut International d'Agriculture, Rome, pp. 237–241.

Anon. (1929) *Actes de la Conférence Internationale pour la Protection des Plantes.* Institut International d'Agriculture, Rome, pp. 189–201.

Anon. (1969) *David Lubin (1849–1919). An Appreciation.* Reprinted from: Proceedings of the Council of the Food and Agriculture Organization, 52nd Session. Food and Agriculture Organization, Rome.

Anon. (1993) *NAKB 50 jaar. Het ontstaan van de NAKB.* Nederlandse Algemene Keuringsdienst voor Boomkwekerijgewassen, Roelofarendsveen, The Netherlands.

Anon. (1995) *Principles of Plant Quarantine as Related to International Trade.* International Standards for Phytosanitary Measures No. 1. Food and Agriculture Organization of the United Nations, Rome.

Anon. (1996a) *Glossary of Phytosanitary Terms.* European and Mediterranean Plant Protection Organisation, Paris.

Anon. (1996b) *Guidelines for Pest Risk Analysis.* International Standards for Phytosanitary Measures No. 2. Food and Agriculture Organization of the United Nations, Rome.

Anon. (1996c) *Import Regulations. Code of Conduct for the Import and Release of Exotic Biological Control Agents.* International Standards for Phytosanitary Measures No. 3. Food and Agriculture Organization of the United Nations, Rome.

Anon. (1996d) *Pest Surveillance. Requirements for the Establishment of Pest Free Areas.* International Standards for Phytosanitary Measures No. 4. Food and Agriculture Organization of the United Nations, Rome.

Anon. (1997a) *Guidelines for Surveillance*. International Standards for Phytosanitary Measures No. 6. Food and Agriculture Organization of the United Nations, Rome.

Anon. (1997b) *Export Certification System*. International Standards for Phytosanitary Measures No. 7. Food and Agriculture Organization of the United Nations, Rome.

Anon. (1998a) *Determination of Pest Status in an Area*. International Standards for Phytosanitary Measures No. 8. Food and Agriculture Organization of the United Nations, Rome.

Anon. (1998b) *Guidelines for Pest Eradication Programmes*. International Standards for Phytosanitary Measures No. 9. Food and Agriculture Organization of the United Nations, Rome.

Anon. (1998c) *Code of Practice for the Management of Agricultural and Horticultural Waste*. Ministry of Agriculture, Fisheries and Food, London.

Anon. (1999a) *International Plant Protection Convention*. New revised text. Food and Agriculture Organization of the United Nations, Rome.

Anon. (1999b) *Requirements for the Establishment of Pest Free Places of Production and Pest Free Production Sites*. International Standards for Phytosanitary Measures No. 10. Food and Agriculture Organization of the United Nations, Rome.

Anon. (2000) *Naktuinbouw at a Glance*. Naktuinbouw, Roelofarendsveen, The Netherlands.

Anon. (2001a) *Pest Risk Analysis for Quarantine Pests*. International Standards for Phytosanitary Measures No. 11. Food and Agriculture Organization of the United Nations, Rome.

Anon. (2001b) *Guidelines for Phytosanitary Certificates*. International Standards for Phytosanitary Measures No. 12. Food and Agriculture Organization of the United Nations, Rome.

Anon. (2001c) *Guidelines for Notification of Non-compliance and Emergency Action*. International Standards for Phytosanitary Measures No. 13. Food and Agriculture Organization of the United Nations, Rome.

Anon. (2002a) *Glossary of Phytosanitary Terms 2002*. International Standards for Phytosanitary Measures No. 5. Food and Agriculture Organization of the United Nations, Rome.

Anon. (2002b) *The Use of Integrated Measures in a Systems Approach for Pest Risk Management*. International Standards for Phytosanitary Measures No. 14. Food and Agriculture Organization of the United Nations, Rome.

Anon. (2002c) *Regulated Non-quarantine Pests: Concept and Application*. International Standards for Phytosanitary Measures No. 16. Food and Agriculture Organization of the United Nations, Rome.

Anon. (2002d) *Pest Reporting*. International Standards for Phytosanitary Measures No. 17. Food and Agriculture Organization of the United Nations, Rome.

Anon. (2003a) *Guidelines for Regulating Wood Packaging Material in International Trade*. International Standards for Phytosanitary Measures No. 15. Food and Agriculture Organization of the United Nations, Rome.

Anon. (2003b) *Guidelines for Regulated Pest Lists*. International Standards for Phytosanitary Measures No. 18. Food and Agriculture Organization of the United Nations, Rome.

Anon. (2003c) *Guidelines for the Use of Irradiation as a Phytosanitary Measure.* International Standards for Phytosanitary Measures No. 19. Food and Agriculture Organization of the United Nations, Rome.

Appel, O. (1915) Leaf roll diseases of the potato. *Phytopathology* 5, 139–148.

Baker, C.R.B., Barker, I., Bartlett, P.W. and Wright, D.M. (1993) Western flower thrips, its introduction and spread in Europe and role as a vector of tomato spotted wilt virus. In: Ebbels, D.L. (ed.) *Plant Health and the European Single Market.* Monograph No. 54, British Crop Protection Council, Farnham Royal, UK, pp. 355–360.

Baker, R.H.A., MacLeod, A. and Sansford, C.E. (1999) Pest risk analysis: the UK experience. *ANPP Fifth International Conference on Pests in Agriculture, Montpellier*, pp. 119–126.

Baker, R.H.A., Sansford, C.E., Jarvis, C.H., Cannon, R.J.C., MacLeod, A. and Walters, K.F.A. (2000) The role of climatic mapping in predicting the potential geographical distribution of non-indigenous pests under current and future climates. *Agriculture, Ecosystems and Environment* 82, 57–71.

Barker, I. (1996) *Serological Methods in Crop Protection.* BCPC Symposium Proceedings No. 65, pp. 13–22.

Barker, K.R. and Davis, E.L. (1996) Assessing plant-nematode infestations and infections. *Advances in Botanical Research* 23, 103–136.

Bartlett, P.W. (1979) Preventing the establishment of Colorado beetle in England and Wales. In: Ebbels, D.L. and King, J.E. (eds) *Plant Health: the Scientific Basis for Administrative Control of Plant Diseases and Pests.* Blackwell Scientific Publications, Oxford, pp. 247–257.

Bartlett, P.W. and Macdonald, O.C. (1993) *Spodoptera littoralis*: experience with an alien polyphagous glasshouse pest in the United Kingdom. In: Ebbels, D.L. (ed.) *Plant Health and the European Single Market.* Monograph No. 54, British Crop Protection Council, Farnham Royal, UK, pp. 371–376.

Biddle, A. (ed.) (2001) *Seed Treatment: Challenges and Opportunities.* BCPC Symposium Proceedings No. 76.

Bourke, P.M.A. (1964) Emergence of potato blight, 1843–46. *Nature* 203, 805–808.

Bridge, P. (2002) Biochemical and molecular techniques. In: Waller, J.M., Lenné, J.M. and Waller, S.J. (eds) *Plant Pathologist's Pocketbook*, 3rd edn. CAB International, Wallingford, UK, pp. 229–244.

Brookes, M. (1998) Running wild. *New Scientist* 160, 38–41.

Butler, E.J. and Jones, S.G. (1949) *Plant Pathology.* MacMillan, London.

Cannon, R.J.C. (1996) *Bacillus thuringensis* use in agriculture: a molecular perspective. *Biological Reviews* 71, 561–636.

Cannon, R.J.C., Pemberton, A.W. and Bartlett, P.W. (1999) Appropriate measures for the eradication of unlisted pests. *EPPO Bulletin* 29, 29–36.

Chambers, D.A. (1985) Field performance of Wye Target hops on 'wilt' land. *East Malling Research Station Report for 1984*, pp. 231–232.

Cheek, S. (1999) Control options for eradication campaigns against quarantine pests: what now? *EPPO Bulletin*, 29, 55–61.

Chock, A.K. (1979) The International Plant Protection Convention. In: Ebbels, D.L. and King, J.E. (eds) *Plant Health: the Scientific Basis for Administrative Control of Plant Diseases and Pests.* Blackwell Scientific Publications, Oxford, pp. 1–11.

Clark, M.F. and Adams, A.N. (1977) Characteristics of the microplate method of enzyme-linked immunosorbent assay for the detection of plant viruses. *Journal of General Virology* 34, 475–483.

Clifford, B.C. and Lester, E. (eds) (1988) *Control of Plant Diseases: Costs and Benefits*. Blackwell Scientific Publications, Oxford.

Collins, D.W. (1996) The separation of *Liriomyza huidobrensis* (Diptera: Agromyzidae) from related indigenous and non-indigenous species encountered in the United Kingdom using cellulose acetate electrophoresis. *Annals of Applied Biology* 128, 387–398.

De Bach, P. and Rosen, D. (1991) *Biological Control by Natural Enemies*, 2nd edn. Cambridge University Press, Cambridge.

Debergh, P.C. and Zimmerman, R.H. (eds) (1991) *Micropropagation: Technology and Application*. Kluwer Academic Publishers, Dordrecht, The Netherlands.

De Boer, S.H., Cuppels, D.A. and Gitaitis, R.D. (1996a) Detecting latent bacterial infections. *Advances in Botanical Research* 23, 27–57.

De Boer, S.H., Slack, S.A., van den Bovenkamp, G.W. and Mastenbroek, I. (1996b) A role for pathogen indexing procedures in potato certification. *Advances in Botanical Research* 23, 217–242.

Dewey, F.M. (2002) Immunological techniques. In: Waller, J.M., Lenné, J.M. and Waller, S.J. (eds) *Plant Pathologist's Pocketbook*, 3rd edn. CAB International, Wallingford, UK, pp. 221–228.

Diamond, J. (1998) *Guns, Germs and Steel. A Short History of Everybody for the Last 13,000 Years*. Vintage Random House UK, London, pp. 104–113.

Dickens, J.S.W. (ed.) (1979) *Diseases of Bulbs*. Her Majesty's Stationery Office, London, pp. 87–88.

Duncan, J. and Cooke, D. (2002) Identifying, diagnosing and detecting *Phytophthora* by molecular methods. *Mycologist* 16, 59–66.

Ebbels, D.L. (1979) A historical review of certification schemes for vegetatively-propagated crops in England and Wales. *ADAS Quarterly Review* 32, 21–58.

Ebbels, D.L. (1988) The costs and benefits of seed potato classification. In: Clifford, B.C. and Lester, E. (eds) *Control of Plant Diseases: Costs and Benefits*. Blackwell Scientific Publications, Oxford, pp. 115–122.

Ebbels, D.L. (1989) Administrative control of pome fruit planting material in countries other than Canada and the United States. In: Fridlund, P.R. (ed.) *Virus and Viruslike Diseases of Pome Fruits and Simulating Noninfectious Disorders*. Cooperative Extension College of Agriculture and Home Economics, Washington State University, Pullman, USA, pp. 314–323.

Ebbels, D.L. (1990) *Plant Health: Statutory Campaigns for Eradication or Control of Plant Diseases*. Report No. 21. ADAS Central Science Laboratory, Harpenden, UK.

Ebbels, D.L. (ed.) (1993). *Plant Health and the European Single Market*. Monograph No. 54, British Crop Protection Council, Farnham Royal, UK.

Ebbels, D.L. and King, J.E. (eds) (1979) *Plant Health: the Scientific Basis for Administrative Control of Plant Diseases and Pests*. Blackwell Scientific Publications, Oxford.

Elphinstone, J.G., Stanford, H. and Stead, D.E. (1998) Detection of *Ralstonia solanacearum* in potato tubers, *Solanum dulcamara* and associated irrigation water. In: Prior, P., Allen, C. and Elphinstone, J.G. (eds) *Bacterial Wilt Disease: Molecular and Ecological Aspects*. Springer-Verlag, Berlin/Heidelberg, pp 133–139.

EPPO (1993a) EPPO Standard PM 5/1(1). Guidelines on pest risk analysis (PRA). Checklist of information required for pest risk analysis (PRA). *EPPO Bulletin* 23, 191–198.

EPPO (1993b) EPPO Standard PM 5/2(1). Guidelines on pest risk analysis (PRA). Pest risk analysis to decide immediate action to be taken on interception of a pest in an EPPO country. *EPPO Bulletin* 23, 199–202.

EPPO (1997) EPPO Standard PM 5/3(1). Guidelines on pest risk analysis. Pest risk assessment scheme. *EPPO Bulletin* 27, 281–305.

Evans, H.F. and Fielding, N.J. (1996) Restoring the natural balance: biological control of *Dendroctonus micans* in Great Britain. In: Waage, J.K. (Chairman) *Biological Control Introductions: Opportunities for Improved Crop Production*. BCPC Symposium Proceedings No. 67, pp. 47–57.

Foster, N. (2000) *Blackstone's Guide to EC Legislation*. Blackstone Press, London.

Fox, R.T.V. (1993) *Principles of Diagnostic Techniques in Plant Pathology*. CAB International, Wallingford, UK.

Fridlund, P.R. (ed.) (1989) *Virus and Viruslike Diseases of Pome Fruits and Simulating Noninfectious Disorders*. Cooperative Extension College of Agriculture and Home Economics, Washington State University, Pullman, USA.

Fry, W.E., Goodwin, S.B., Dyer, A.T., Matuszak, J.M., Drenth, A., Tooley, P.W., Sujkowski, L.S., Koh, Y.J., Cohen, B.A., Spielman, L.J., Deahl, K.L., Inglis, D.A. and Sandlan, K.P. (1993) Historical and recent migrations of *Phytophthora infestans*: chronology, pathways and implications. *Plant Disease* 77, 653–661.

Fulling, E.H. (1942) Plant life and the law of man. III. Barberry eradication. *Journal of the New York Botanical Garden* 43, 152–157.

Fulling, E.H. (1943) Plant life and the law of man. IV. Barberry, currant and gooseberry, and cedar control. *The Botanical Review* 9, 483–592.

Griffin, R.L. (2000) The precautionary approach and phytosanitary measures. *Proceedings of the Brighton Crop Protection Conference – Pests and Diseases – 2000*. British Crop Protection Council, Farnham Royal, UK, Vol. 3, pp. 1153–1158.

Hallman, G.J. (1998) Ionizing radiation quarantine treatments. *Anais da Sociedade Entomologica do Brasil* 27, 313–323.

Hartmann, H.T., Kester, D.E. and Davies, F.T., Jr (1990) *Plant Propagation; Principles and Practices*, 5th edn. Prentice-Hall International, Englewood Cliffs, New Jersey, pp. 437–441.

Hewett, P.D. (1979) Regulating seed-borne disease by certification. In: Ebbels, D.L. and King, J.E. (eds) *Plant Health: the Scientific Basis for Administrative Control of Plant Diseases and Pests*. Blackwell Scientific Publications, Oxford, pp. 163–173.

Hewitt, W.B. and Chiarappa, L. (eds) (1977) *Plant Health and Quarantine in International Transfer of Genetic Resources*. CRC Press, Cleveland, Ohio.

Hirst, J.M., Hide, G.A., Griffith, R.L. and Stedman, O.J. (1970) Improving the

health of seed potatoes. *Journal of the Royal Agricultural Society* 131, 87–106.
Hobhouse, H. (1992) *Seeds of Change. Five Plants that Transformed Mankind.* PaperMac, London.
Hokkanen, H.M.T. and Lynch, J.M. (eds) (1995) *Biological Control: Benefits and Risks.* Cambridge University Press, Cambridge.
Holliday, P. (1998) *A Dictionary of Plant Pathology*, 2nd edn. Cambridge University Press, Cambridge, UK.
Hollings, M. (1965) Disease control through virus-free stock. *Annual Review of Phytopathology* 3, 367–396.
Hopper, B.E. (1995) *NAPPO Compendium of Phytosanitary Terms.* North American Plant Protection Organisation, Nepean, Ontario.
Hutchins, J.D. and Reeves, J.C. (eds) (1997) *Seed Health Testing: Progress Towards the 21st Century.* CAB International, Wallingford, UK.
Irwin, M.P.S. (1989) The genus *Quelea*. In: Mundy, P.J. and Jarvis, M.J.F. (eds) *Africa's Feathered Locust.* Baobab Books, Academic Books, Harare, Zimbabwe, pp. 9–13.
Jones, P. (2002) Phytoplasma plant pathogens. In: Waller, J.M., Lenné, J.M. and Waller, S.J. (eds) *Plant Pathologist's Pocketbook,* 3rd edn. CAB International, Wallingford, UK, pp. 126–139.
Kahn, R.P. (ed.) (1989) *Plant Protection and Quarantine*, Vols 1–3. CRC Press, Boca Raton, Florida.
Kahn, R.P. and Mathur, S.B. (eds) (1999) *Containment Facilities and Safeguards for Exotic Plant Pathogens and Pests.* APS Press, The American Phytopathological Society, St Paul, Minnesota.
Katan, J. (1981) Solar heating (solarization) of soil for control of soilborne pests. *Annual Review of Phytopathology* 19, 211–236.
Koeman, J.H. and Zadoks, J.C. (1999) History and future of plant protection policy, from ancient times to WTO-SPS. In: Meester, G., Woittiez, R.D. and de Zeeuw, A. (eds) *Plants and Politics.* Wageningen Press, Wageningen, The Netherlands, pp. 21–48.
Krall, S., Pevelin, R. and Badiallo, D. (1997) *New Strategies in Locust Control.* Birkhäuser-Verlag, Basel, Switzerland.
Large, E.C. (1940) *The Advance of the Fungi.* Jonathan Cape, London.
Leonard, D. (2000) *The Economist Guide to the European Union*, 6th edn. The Economist, London.
Lévesque, C.A. and Eaves, D.M. (1996) A decision modelling approach for quantifying risk in pathogen indexing. *Advances in Botanical Research* 23, 243–277.
Ling, L. (1953) International Plant Protection Convention: its history, objectives and present status. *FAO Plant Protection Bulletin* 1, 65–68.
Lomer, C. and Prior, C. (eds) (1992) *Biological Control of Locusts and Grasshoppers.* CAB International, Wallingford, UK.
MacKenzie, D. (2001) All fall down. *New Scientist*, 172, 34-37.
MacKenzie, D.R., Marchetti, M.A. and Kingsolver, C.H. (1985) The potential for the willful introduction of biotic agents as an act of anticrop warfare. In: MacKenzie, D.R., Barfield, C.S., Kennedy, G.G., Berger, R.D. and Taranto, D.J. (eds) *The Movement and Dispersal of Agriculturally Important Biotic Agents.* Claitor's Publishing Division, Baton Rouge, Louisiana, pp. 601–608.

McNamara, D.G. and Cleia, M.F.B. (1985) Pests and diseases of strawberries. *East Malling Research Station Report for 1984*, pp. 161–162.

McRae, C.F. and Wilson, D. (2002) Plant health as a trade policy issue. *Australasian Plant Pathology* 31, 103–105.

Marshall, G. (Chairman) (1996) *Diagnostics in Crop Production*. BCPC Symposium Proceedings No. 65.

Martelli, G.P. (ed.) (1992) *Grapevine Viruses and Certification in EEC Countries: State of the Art*. Centre International de Hautes Etudes Agronomiques Mediterraneennes and Istituto Agronomico Mediterraneo, Bari, Italy.

Mears, D.R. and Khan, R.P. (1999) New concepts and designs for quarantine greenhouses. In: Kahn, R.P. and Mathur, S.B. (eds) *Containment Facilities and Safeguards for Exotic Plant Pathogens and Pests*. APS Press, The American Phytopathological Society, St Paul, Minnesota, pp. 79–92.

Meinzingen, W.F. (ed.) (1993) *A Guide to Migrant Pest Management in Africa*. Food and Agriculture Organization of the United Nations, Rome.

Miller, S.A. (1996) Detecting propagules of plant pathogenic fungi. *Advances in Botanical Research* 23, 73–102.

Mills, P.R. (1996) *Use of Molecular Techniques for the Detection and Diagnosis of Plant Pathogens*. BCPC Symposium Proceedings No. 65, pp. 23–32.

Mitchell, W.C. and Saul, S.H. (1990) Current control methods for the Mediterranean fruit fly, *Ceratitis capitata*, and their application in the USA. *Review of Agricultural Entomology* 78, 923–940.

Mumford, R., Walsh, K., Barker, I. and Boonham, N. (1999) Fluorescent PCR for the detection of potato viruses. *Petria* 9, 127–130.

Mumford, R.A., Walsh, K. and Boonham, N. (2000) Comparison of molecular methods for the routine detection of viroids. *EPPO Bulletin* 30, 431–435.

Mundy, P.J. and Jarvis, M.J.F. (eds) (1989) *Africa's Feathered Locust*. Baobab Books, Academic Books, Harare, Zimbabwe.

Navi, S.S. and Bandyopadhyay, R. (2002) Biological control of fungal plant pathogens. In: Waller, J.M., Lenné, J.M. and Waller, S.J. (eds) *Plant Pathologist's Pocketbook*, 3rd edn. CAB International, Wallingford, UK, pp 354–365.

Nkouka, N. (1992) IAPSC/OAU communiqué from the Inter-African Scientific Secretariat of the OAU, with comments on the application of the Inter-African Plant Protection Convention. In: Lomer, C. and Prior, C. (eds) *Biological Control of Locusts and Grasshoppers*. CAB International, Wallingford, UK, pp. 50–53.

Orton, W.A. (1914) Inspection and certification of potato seed stock. *Phytopathology* 4, 39–40.

Parsons, F.G., Garrison, C.S. and Beeson, K.E. (1961) Seed certification in the United States. *Seeds: the Yearbook of Agriculture 1961*. United States Department of Agriculture, Washington, DC, pp. 394–401.

Pemberton, A.W. (1988) Quarantine: the use of cost:benefit analysis in the development of MAFF plant health policy. In: Clifford, B.C. and Lester, E. (eds) *Control of Plant Diseases: Costs and Benefits*. Blackwell Scientific Publications, Oxford, pp. 195–202.

Posnette, A.F. (1953) Heat inactivation of strawberry viruses. *Nature* 171, 312.

Pratt, M.A. (1979) Potato wart disease and its legislative control in England and Wales. In: Ebbels, D.L. and King, J.E. (eds) *Plant Health: the Scientific Basis for Administrative Control of Plant Diseases and Pests.* Blackwell Scientific Publications, Oxford, pp. 199–212.

Ramsey, A. (ed.) (2000) *Eurojargon: a Dictionary of European Union Acronyms, Abbreviations and Sobriquets,* 6th edn. CPI Ltd, Bruton, UK.

Reed, P.J. and Dickens, J.S.W. (1993) Evaluation of various disinfectants against strawberry red core (*Phytophthora fragariae*). Tests of Agrochemicals and Cultivars 14. *Annals of Applied Biology* 122 (suppl.), 54–55.

Rieman, G.H. (ed.) (1956) Early history of potato seed certification in North America, 1913–1922. *Potato Handbook.* The Potato Association of America, New Brunswick, pp. 6–10.

Riley, C.V. (1877) *The Colorado Beetle, With Suggestions for its Repression and Methods of Destruction.* George Routledge and Sons, London.

Rogers, A.G. (1914) The international phytopathological conference, 1914. *Annals of Applied Biology* 1, 113–117.

Rosenberg, D.Y. and Aichele, M.D. (1989) Virus-free certification programs in the United States and Canada. In: Fridlund, P.R. (ed.) *Virus and Viruslike Diseases of Pome Fruits and Simulating Noninfectious Disorders.* Cooperative Extension College of Agriculture and Home Economics, Washington State University, Pullman, USA, pp. 299–307.

Rudd-Jones, D. and Langton, F.A. (eds) (1986) *Healthy Planting Material: Strategies and Technologies.* Monograph No. 33, British Crop Protection Council, Farnham Royal, UK.

Salmon, E.S. (1914) Observations on the life-history of the American gooseberry-mildew (*Sphaerotheca mors-uvae* (Schwein.) Berk.). *Annals of Applied Biology* 1, 177–182.

Sansford, C.E. (1998) Karnal bunt (*Tilletia indica*): detection of *Tilletia indica* Mitra in the US: potential risk to the UK and the EU. In: Malik, V.S. and Mathre, D.E. (eds) *Bunts and Smuts of Wheat: an International Symposium; North Carolina, August 17–20, 1997.* NAPPO, Ottawa, pp. 273–302.

Shattock, R.C. and Day, J.P. (1996) Migration and displacement; recombinants and relicts: 20 years in the life of potato late blight (*Phytophthora infestans*). *Proceedings of the Brighton Crop Protection Conference.* British Crop Protection Council, Farnham Royal, UK, Vol. 3, pp. 1129–1136.

Shepard, J.F. and Claflin, L.E. (1975) Critical analyses of the principles of seed potato certification. *Annual Review of Phytopathology* 13, 271–293.

Skerritt, J.H. and Appels, R. (eds) (1995) *New Diagnostics in Crop Sciences.* CAB International, Wallingford, UK.

Smith, I.M. (1979) EPPO: the work of a regional plant protection organization, with particular reference to phytosanitary regulations. In: Ebbels, D.L. and King, J.E. (eds) *Plant Health: the Scientific Basis for Administrative Control of Plant Diseases and Pests.* Blackwell Scientific Publications, Oxford, pp. 13–22.

Smith, I.M., McNamara, D.G., Scott, P.R. and Holderness, M. (eds) (1996) *Quarantine Pests for Europe,* 2nd edn. CAB International, Wallingford, UK, and the European and Mediterranean Plant Protection Organisation, Paris, France.

Sommer, N.F. and Mitchell, F.G. (1986) Gamma irradiation – a quaratine treatment for fresh fruits and vegetables. *Horticultural Science* 21, 356–360.

Spielman, L.J., Drenth, A., Davidse, L.C., Sujkowski, L.J., Gu, W., Tooley, P.W.

and Fry, W.E. (1991) A second world-wide migration and population displacement of *Phytophthora infestans*? *Plant Pathology* 40, 422–430.

Stead, D.E. (1995) Profiling techniques for the identification and classification of plant pathogenic bacteria. *EPPO Bulletin* 25, 143–150.

Steedman, A. (1990) *Locust Handbook*, 3rd edn. Natural Resources Institute, Chatham, UK.

Thompson, J.R. (1974) The International Seed Testing Association 1924–74. *Seed Science and Technology* 2, 267–283.

Traynor, P.L., Adair, D. and Irwin, R. (2001) *A Practical Guide to Containment: Greenhouse Research with Transgenic Plants and Microbes*. Information Systems for Biotechnology, Blacksburg, Virginia, USA.

UNEP (1992) *Convention on Biological Diversity 22 May 1992*. UNEP, Nairobi, Kenya.

UNEP (2000) *Cartagena Protocol on Biosafety to the Convention on Biological Diversity, 29 January 2000*. UNEP, CBD, Montreal, Canada.

UNEP (2002) *Alien Species: Guiding Principles for the Prevention, Introduction and Mitigation of Impacts*. Report to the 6th meeting of the Conference of Parties to the Convention on Biological Diversity 7–19 April 2002, The Hague, The Netherlands.

Usinger, R.L. (1964) The role of Linnaeus in the advancement of entomology. *Annual Review of Entomology* 9, 1–16.

Van der Plank, J.E. (1963) *Plant Diseases: Epidemics and Control*. Academic Press, London, pp. 212–221.

Voller, A., Bartlett, A., Bidwell, D.E., Clark, M.F. and Adams, A.N. (1976) The detection of viruses by enzyme-linked immunosorbent assay (ELISA). *Journal of General Virology* 33, 165–167.

Waller, J.M. (2002) Virus diseases. Revised by J.M. Waller. In: Waller, J.M., Lenné, J.M. and Waller, S.J. (eds) *Plant Pathologist's Pocketbook,* 3rd edn. CAB International, Wallingford, UK, pp. 108–125.

Waller, J.M. and Cannon, P.F. (2002) Fungi as plant pathogens. In: Waller, J.M., Lenné, J.M. and Waller, S.J. (eds). *Plant Pathologist's Pocketbook*, 3rd edn. CAB International, Wallingford, UK, pp. 75–93.

Ward, P. (1971) The migration patterns of *Quelea quelea* in Africa. *Ibis* 113, 275–297.

Wilhelm, S. (1966) Chemical treatments and inoculum potential of soil. *Annual Review of Phytopathology* 4, 53–78.

Williams, C.B. (1948) The Rothamsted light trap. *Proceedings of the Royal Entomological Society of London A*. 23, 80–85.

Woiwod, I.P. and Harrington, R. (1994) Flying in the face of change: the Rothamsted Insect Survey. In: Leigh, R.A. and Johnson, A.E. (eds) *Long-term Experiments in Agricultural and Ecological Sciences*. CAB International, Wallingford, UK, pp. 321–342.

Zimmerman, R.H., Griesbach, R.J., Hammerschlag, F.A. and Lawson, R.H. (1986) *Tissue Culture as a Plant Production System for Horticultural Crops*. Martinus Nijhoff Publishers, Lancaster, UK.

Websites

Australia Group: www.australiagroup.net/intro.htm
Biological Weapons Convention (BWC): www.acronym.org.uk/bwc/
Biosafety levels, USA National Institutes of Health: www4.od.nih.gov/
Comité Regiónal de Sanidad Vegetal para el Cono Sur (COSAVE): www.cosave.org.py/baseesp.htm
Comunidad Andina (CA): www.comunidadandina.org/
Convention on Biological Diversity (CBD): www.biodiv.org/meetings/final-reports.asp
Convention on Biological Diversity (CBD): www.biodiv.org/chm/conv/default.htm
Convention on Biological Diversity, Biosafety Protocol: www.biodiv.org/biosafe/protocol/Protocol.html
Convention on Biological Diversity, Guiding Principles on Alien Species: www.biodiv.org/doc/meetings/cop/cop-06/official/cop-06-20-en-pdf (pp. 247–252)
European and Mediterranean Plant Protection Organisation (EPPO): www.eppo.org/
European Union (EU): www.europa.eu.int
European Union, index, information on plant health: www.europa.eu.int/comm/food and www.euro-know.org/dictionary/
Food and Agriculture Organization of the United Nations (FAO): www.fao.org/
Information Systems for Biotechnology: www.isb.vt.edu/
International Plant Protection Convention (IPPC), Secretariat, FAO: www.ippc.int/
International Standards for Phytosanitary Measures (ISPMs), International Plant Protection Convention, International Phytosanitary Portal, FAO: www.ippc.int/cds_ippc/IPP/En/default.htm
Methyl bromide research, United States Department of Agriculture, Agricultural Research Service: www.ars.usda.gov/is/mb/mebrweb.htm
Methyl bromide, national response strategy, Environment Australia: www.ea.gov.au/atmosphere/ozone/
Migrant Pests Group, FAO: www.fao.org/news/global/locusts/locuhome.htm
Montreal Protocol on Substances that deplete the Ozone Layer, United Nations Environment Programme: www.unep.org/ozone/vienna.shtml
Montreal Protocol, methyl bromide information, including the Critical Use Exemption Process, United States Environmental Protection Agency: www.epa.gov/ozone/mbr
North American Free Trade Agreement (NAFTA), Secretariat: www.nafta-sec-alena.org/
North American Free Trade Agreement, Sanitary and Phytosanitary Measures: www.sice.oas.org/trade/nafta/chap-073.asp#A709
North American Plant Protection Organization: www.nappo.org/
Organization for Economic Co-operation and Development (OECD), certification schemes: www.oecd.org/agr/code/
Organismo Internacional Regional de Sanidad Agropecuaria: http://ns1.oirsa.org.sv/
Pacific Plant Protection Organization (PPPO): www.spc.org.nc/

Regional Plant Protection Organizations: www.ippc.int/cds_ippc/IPP/En/default.htm
Sunshine Project: www.sunshine-project.org
United Nations Economic Commission for Europe (ECE): www.unece.org
World Trade Organization (WTO): www.wto.org

Index

Acrididae 137
agreements
  bilateral 46–47
  international 13, 45
  regional 45
Agrobacterium tumefaciens 102, 157
Albugo spp. 187
Allium
  cepa var. *ascallonicum* 179
  fistulosum 179
  sativum 179
Anti-locust Research Centre 141
Appel, Otto 19
Application of Sanitary and Phytosanitary Measures, Agreement on *see* WTO-SPS
APPPC 36, 252–253
archives *see* record-keeping
armyworms 139, 140–142
Ascochyta spp. 65
Asia and Pacific Plant Protection Commission *see* APPPC
Australia 137, 139
Australia Group 102, 271–274
  awareness-raising guidelines 272, 273
  common control lists 273
  export controls 272–274

Bacillus thuringiensis var. *kurstaki* 130
Bactrocera
  cucurbitae 132
  tryoni 225
Barbados 253
Beauveria bassiana 138
Beet necrotic yellow vein virus 188, 230
Berberis vulgaris 9, 10, 11
biological control *see* phytosanitary action measures
biological weapons 268–275
  information on 275
Biological Weapons Convention 1972 *see* BWC
Biosafety Protocol 43
BIPs 80–83, 103
BKD 24, 169
Bloembollenkeuringsdienst 24, 169
bonsai 97
border entry points 103
border inspection posts *see* BIPs
Bos, Ritzema 22
BWC 47, 270–271, 272, 274, 275

CA 36, 255

293

Cactoblastis cactorum 130
Calliptamus italicus 137
Cameroon 257
Canada 17, 19, 31, 171, 259
Caribbean Plant Protection
    Commission *see* CPPC
Carter, W. 237
Cassytha 189
Catharanthus roseus 1 89
CBD 29, 43–44, 269
    Cartegena Protocol on Biosafety
        43
    Guiding Principles on Alien
        Species 43, 44
Centraal Comité voor de Keuring van
        Gewassen in Nederland 20
Central Committee for the Inspection
        of Plants in The Netherlands
        20
Centro Internacional de la Papa 167
Ceratitis capitata 132, 224
certification schemes
    botanical purity 151, 162–164
    clonal varieties 152
    Community grades 74, 175
    compulsory 168–169
    cost:benefit of 169
    crop inspections in 156, 157, 167
    crop vigour 153, 164
    crops covered 150, 161
    development 18–20, 24, 26, 27,
        145, 148
    ECE 171
    EPPO 171–174
    health 153, 164
    keeping up-to-date 170
    laboratory tests 166, 210
    literature on 44
    measures to control pathogen
        spread 154–158, 162, 165
    OECD 177, 182
    pathogens amenable to control
        by 146, 153
    pesticides in 155
    physiological quality 164
    preventing cross-pollination in
        154, 157, 163
    procedures 149
    seeds 20, 21, 161–167
    standards 149
    terminology 144, 145, 174
    tissue cultured material 151, 160
    tolerances 149, 152
    undesirable variations in 152
    vegetatively propagated material
        150–158
    voluntary 168–169
Channel Islands 85
Chenopodium
    amaranticolor 206
    quinoa 206
Chortoicetes terminifera 137
Chromista 184
CIP 167
CIRSA 261
classification schemes 173–174
Clavibacter michiganensis ssp.
        sepedonicus 119, 175
Claviceps spp. 165
    purpurea 65
CLIMEX 222, 225
Coccinella spp. 12
Cochliomyia hominivorax 132
Codex Alimentarius Commission 40
Colletotrichum 208
Colorado beetle *see Leptinotarsa*
    decemlineata
Comité Internacional Regional de
        Sanidad Agropecuaria 261
Comité Permanent Inter-Etats de
        Lutte contre la Sécheresse
        dans le Sahel 142
Comité Regional de Sanidad Vegetal
        del Cono Sur *see* COSAVE
communications 113
compensation, financial 68, 135–136
Comunidad Andina *see* CA
congresses, international 21–22, 23
containment facilities 244–250
    plant material 245–249
    plant pests 249–250
contingency plans 126
Convention on Biological Diversity
    see CBD
Convention, Biological Weapons *see*
    BWC
Convention, International Plant
    Protection *see* IPPC

Convention, International, for the
    Protection of Plants 1929 23,
    30
Convention on *Phylloxera* 1878
    263–267
COREPER 53
Corynebacterium fascians 157
COSAVE 36, 46, 254–255
cost:benefit analysis 134
CPPC 36, 253
Cronartium ribicola 11
crop variety identification 81, 151,
    162, 208
Cucumis sativus 206
culture collections 102–103, 274
Cuscuta 189
customs services, liaison with 103
Cynara scolymus 179

De Bary, Anton 9
Delia antiqua 132
Dendroctonus micans 130
Denmark 10, 11, 21, 161, 169
Desert Locust Control Organization
    for Eastern Africa 142
Desert Locust Information Service
    142
Directive 2000/29/EC 45, 50, 56,
    60–67, 71, 118
Directive 77/93/EEC 49, 60
disinfectants 82, 234–236
Dociostaurus maroccanus 137
dodder, transmission by 189
dunnage 97
Dutch Flower Bulb Inspection
    Service 24
Dutch Inspection Service for Tree
    Crops 25
Dutch Pomological Association 19

East Malling Research Station 26
EEC 48
Eisenia
    andrei 242
    fetida 242
El Salvador 260
Encarsia formosa 130

enzyme-linked immunosorbent assay
    195–197
Epidinocarsis lopezi 130
EPPO 28, 35, 36, 63, 90, 179, 209,
    213, 222, 233, 238, 252,
    255–257
EPPO schemes 172–174
Erkelens, Mr 25
Erwinia
    amylovora 110, 235
    carotovora 175, 247
Ethiopia 142
European and Mediterranean Plant
    Protection Organization *see*
    EPPO
European Atomic Energy Community
    48
European Coal and Steel Community
    48
European Community 49, 50
    legislation *see* legislation, EC
European Economic Community *see*
    EEC
European Union
    Commission of the 30, 52–53,
        117
    Common Catalogue of varieties
        176, 179–180
    comparative trials 176
    Council of Ministers 51–52
    Council 52
    Court of Justice 54
    Economic and Social Committee
        53
    institutions 51–54
    legislation *see* legislation, EC
    literature on 51
    marketing schemes 176–179
    national lists of varieties 151,
        176, 179–180
        DUS characteristics 180
        VCU characteristics 180
    Parliament 53
    permanent representation 53
    Phytosanitary Inspectorate 56,
        68
        vademecum 68
    plant variety rights 180–182
        farmers' privilege 182

European Union – *continued*
  plant variety rights – *continued*
    variety denomination 181
  Presidency 51
  schemes 174–179
    CAC grade 179
    seed potato certification
      174–176
    standing committees 55–60
    voting 58–60
EUROPHYT 66
exports 107–113
  controls 102
  grain 111
  operations 110
  pre-clearance of 112
  re-exports 111, 119

FAO 29–32, 89, 108, 141, 142, 238, 252
Fiji 262
Food and Agriculture Organization of the United Nations *see* FAO
France 1, 24, 28, 256
Frankliniella occidentalis 220
fumigation 122, 232, 236–239
fungi, classification of 184
Fusarium spp. 165, 215
  oxysporum f.sp. *vasinfectum* 133

GATS 40
GATT 40
General Agreement on Tariffs and Trade 40
General Agreement on Trade in Services 40
Geneva Protocol on Standardization of Fruits and Vegetables 171
geographical information systems *see* GIS
Germany 15, 19, 21, 145, 161, 169, 275
GIS 220, 222–227
  climate mapping 222–223
  climographs 223–224
  indices 224–225

  phenology models 226
  value of 226–227
Globodera
  pallida 175
  rostochiensis 175
GMOs 45, 102, 245
Gomphrena globosa 206
good laboratory practice 210, 211
good plant health practice 228–234
  record keeping in 233–234
good plant protection practice 233
grasshoppers 137–138
grey literature 75
Guiding Principles on Alien Species 43
Güssow, H.T. 19
Gymnosporangium juniperi-virginianae 12

hop *see Humulus lupulus*
Humulus lupulus 25, 134, 150
Hungary 10
hygiene
  disinfectants 234–236
  field crops 229
  protected environments 230–232
    cleaning and disinfection 231–232
  protective clothing 234

IAPSC 36, 257–259
Icerya purchasi 130
imports 89–107
  licences 98–99, 101
  packing materials 97
  permits 98–99
  potato 97
  precautionary measures 89, 94–96
  regulations 93
  tissue culture 97
  wild plants 100
in transit, goods 106–107
  inspections of 107
indexing 184
  uses of 208, 209
India 137, 226

Inspection Institute for Seed and
    Planting Materials 20
inspections *see* phytosanitary
    inspections
Inter-African Phytosanitary Council
    see IAPSC
International Convention for the
    Protection of Plants 1929 23,
    30
International Institute of Agriculture
    22–23, 30
International Office of Epizootics 40
International Plant Protection
    Convention *see* IPPC
International Potato Center 167
International Red Locust Control
    Organization for Central and
    South Africa 142
International Seed Testing
    Association *see* ISTA
International Standards for
    Phytosanitary Measures *see*
    ISPMs
International Union for the
    Protection of New Varieties of
    Plants *see* UPOV
inter-tropical wind convergence zone
    137
IPPC 28, 31–34, 66, 70, 71, 88, 89,
    92, 94, 103, 107, 117, 213,
    222, 252
Ireland 68
    famine in 12
Isle of Man 85
ISPMs 29, 37–39
ISTA 21, 166, 167
Italy 22, 224

Japan 132
Juniperus virginiana 12

Keuringsinstituut voor Zaaizaad en
    Pootgoed 20

laboratories, diagnostic, quality
    assurance in 210

record keeping 210–211
legislation
    barberry destruction 9–11
        Austria 10
        Bavaria 10
        Denmark 10, 11
        France 9
        Hungary 10
        Norway 10
        Prussia 10
        UK 10
        USA 9, 10
    EC phytosanitary 45, 48, 49,
        54–67, 174–179
        bans and prohibitions 63
        derogations 66
        emergency procedures 66–67
        financial compensation 68
        plant passports 50, 64–65
        protected zones 64–65, 67
        registration of producers 65
        responsible official body 64
        third countries 67
        transparency 63
    phytosanitary 72–73
    UK phytosanitary 16, 18, 23–27,
        86
Leptinotarsa decemlineata 14–16,
    17, 24, 27, 28, 115
Lettuce mosaic virus 165
Linnaeus, Carl 12
Liriomyza leaf-miners 121, 208
literature, plant health 3
LMOs *see* GMOs
Locusta migratoria 137
Locustana pardalina 137
locusts 137–139
    international cooperation on
        141–142
Long Ashton Research Station 26
Lubin, David 22

Madagascar 137
Mahonia spp. 11
male annihilation technique 132
Manihot esculenta 150
marketing schemes 153, 176–179
    field crop seeds 176

marketing schemes – *continued*
  fruit plants and propagating
    material 177–179
  ornamentals 177–178
  vegetable plants 177–179
Mauritania 137
medfly *see Ceratitis capitata*
MERCOSUR 46, 254
Metarhizium flavoviride 130, 138
methyl bromide 237–239
Mexico 12
micropropagation 98
Microsphaera grossulariae 16
monitoring *see* pest, surveillance
Montreal Protocol, The 238–239
Murphy, Paul A. 19
Mythimna
  separata 139
  unipuncta 139

NAK 20, 25, 169
NAKB 25
Naktuinbouw 169
NAPPO 35–36, 46, 209, 259–260
narcotic crops, biological control of
  274–275
national listing system 151, 179–180
Nederlandse Algemene
  Keuringsdienst voor
  Boomwekerijgewassen 25
Nederlandse Algemene
  Keuringsdienst voor Zaaizaad
  en Pootgoed van
  Landbouwgewasssen *see*
  NAK
Nederlandse Pomologische
  Vereniging 19
Netherlands General Inspection
  Service *see* NAK
Netherlands Plant Protection Service
  21
Netherlands, The 19, 20, 21, 24, 25,
  31, 132, 169
Nicotiana
  clevelandii 206
  debneyi 206
  glutinosa 206
  tabacum 206

Nomadacris septemfasciata 137
non-compliance
  emergency action on 105, 117
  notification 117
North American Free Trade
  Agreement 46
North American Plant Protection
  Organization *see* NAPPO
Norway 10
nuclear stock 146–148, 208
  containment of 245, 248–249
  pathogen-tested 146
  testing methods 147, 150, 207,
    209
nucleic acid probes 201
nucleic acids 200–206

OAU 258
OCLALAV 142
OECD 177, 182, 271
Oedaleus senegalensis 137
official seed testing stations 166, 177
OIE 40
OIRSA 36, 260–261
Oospora 145
Operophtera brumata 12
Opuntia spp. 130
Organisation Commune de Lutte
  Anti-Acridienne 142
Organisation Commune de Lutte
  Anti-Aviaire 142
Organismo Internacional Regional de
  Sanidad Agropecuaria *see*
  OIRSA
Organization of African Unity 258
Orton, W.A. 19, 145

Pacific Plant Protection Organization
  see PPPO
parasitic plants 2
passports, plant 50, 64–65
pathogen-tested planting material
  26, 146
PCR 201–205
penjing 97
Pennisetum purpureum 141
Peru 255

pest
    classification 89
    detection 114
      bacteria 187
      fungi 187
      invertebrates 186–187
      methods 184–189
      phytoplasmas 187
      viroids 189
      viruses 189
    dispersal 4–5, 7, 11, 12, 14
    interception 115, 120–121
    lists 89, 211, 92, 93
    migratory 136–142
      armyworm 139–140
      bird 140–141
      locust and grasshopper 136–139
    notification 115
    outbreaks 125
    quality 92
    quarantine 90
    regulated non-quarantine 90
    risk analysis see PRA
    risk management 221
      level of phytosanitary protection 41, 42, 221
    specimens 116, 211, 120
    surveillance 92–93, 127–129
    taxonomy 183
    unlisted 91, 118
pesticide haptens 199
pesticides 84, 122, 138, 232–233
pests, as offensive agents 268–274
pests, testing methods for 189
    bioassays 206–207
    biochemical 207–208
      enzyme profiling 207, 208
      fatty acid profiling 207, 208
      nutritional profiling 207
      protein profiling 207, 208
    microscopy
      electron 192–193
      light 190–192
    molecular 200–206
      fluorogenic 5′-nuclease assay 204–205
      nested primers 203, 204
      nucleic acid probes 201
      polymerase chain reaction 201–205
      RFLP 205, 206
      RT-PCR 203
    serological 193–200
      DAS-ELISA 195, 196
      ELISA 195–197
      immunoassay kits 199, 200
      immuno-electron microscopy 197
      immunofluorescence 198
      indirect DAS-ELISA 196, 197
      lateral flow devices 199, 200
    visual inspection 190–192
Phaseolus vulgaris 206
Phoma spp. 165
Phylloxera vastatrix 13, 30, 34
    International Convention on, 1878 263–267
Physalis peruviana 206
Phytopathological Committee, International 21
Phytophthora 184, 204, 208
    fragariae 133, 242
    infestans 12, 230
    A2 mating type 13, 216
phytosanitary action measures 105, 114, 115, 117–134
    administrative 123, 124
    biological control 12, 130–132, 138
    chemical 122–123, 125
    cropping control 132–134
    destruction 124–126
    hygiene 125
    irradiation 123
    longer-term 126–134
    physical 121, 123
    precautionary 118
    for protected environments 129
    short-term 119–126
phytosanitary campaigns 75, 127, 136
    diagnosis 75
    surveys 75, 82
phytosanitary certificate, international 49, 66, 88, 107, 110
    absence of 119

phytosanitary costs and charges 85
phytosanitary enforcement 77–78
phytosanitary inspections 77
    delegated 111
    import 115
    location of 104
    pre-export 111
    targeting 116
phytosanitary inspectors 16, 77
    communications 78
    equipment and facilities 81
    health and safety 84
    instructions 79
    protective clothing 82
    training 80, 81
phytosanitary operations 103
phytosanitary regulations 106
    foreign countries' 107, 108
    format 108, 109
phytosanitary systems 73
    policy and administration 73
    scientific advice and research 74, 76
Phytoseiulus persimilis 130
Pinus strobus 11
Pistacia vera 179
Plantenziektenkundige Dienst 22
Plum pox virus 124
Polymyxa 184
    betae 188, 230
Potato leafroll ilarvirus 155
Potato mop top virus 157
Potato virus X 158, 198
Potato virus Y 156, 198
potato, *Solanum* 12, 17, 18, 69, 97, 116 145, 150, 167, 230, 242, 247
    true seed of 167
    variety nomenclature 19, 20
potatoes seed, certification of 145, 169, 171,172, 174–176, 186, 210
PPPO 36, 262
PRA 39, 76, 96, 116, 118, 126, 127, 134, 212–222
    evaluation 218–221
    information for 214–217
      handling 217
    initiation 314
    preliminary 222

    socio-economic impact 216, 217, 220, 221
precautionary approach 43, 45, 222
propagation of plants, rapid methods for 158–161
    bulbs 160–161
      chipping 160
      scooping 161
      twin-scaling 160
    meristem culture 159
    micropropagation 159
    tissue culture 158–160
Protozoa 184
Pseudomonas syringae pvv. 165
Pseudoperonospora cubensis 17
Puccinia
    glumarum 268
    graminis 9, 10, 268
    striiformis 268
Pyricularia oryzae 268
Pythium 184

Quadraspidiotus perniciosus 22
quality assurance for planting material 19, 20, 143–182
quarantine 1, 120
    containment facilities 101, 245–250
    post-entry 106, 120
    units 245–248
Quelea 142
    cardinalis 140
    erythrops 140
    quelea 140

Ralstonia solanacearum 175, 230
record-keeping
    diagnostic laboratories 210
    good plant health practice 233–234
    phytosanitary authorities 111, 133, 250–251
Regional Plant Protection Organizations *see* RPPOs
Rheum 179
Rhizoctonia 145
Rhizophagus grandis 130

Ribes spp. 11
   uva-crispa 16
Riley, Charles V. 15
Rodolia cardinalis 130
roguing 154
RPPOs 28, 29, 32, 35, 70, 89, 90, 92, 103, 108, 117, 252–262
   map of 36
Russia 16

Say, Thomas 15
Schistocera
   cancellata 137
   gregaria 137
screw-worm fly *see Cochliomyia hominivorax*
seed testing 20–21, 166–167
seed treatment 162, 165, 166
Septoria spp. 165
Single Market, EU 48
soil
   imports of 96, 101
   large-scale movement 250
   sampling 82
Solanum tuberosum see potato, *Solanum*
SOPs 79
Soviet Union 270
Sphaerotheca mors-uvae 16, 17
Spodoptera
   exempta 139
   exigua 139
   littoralis 123
Spongospora 184
   subterranea 157
Standard Operating Procedures 79
Standing Committee
   Agricultural, Horticultural and Forestry Seeds and Plants 57, 175
   Plant Health 55–57, 68
   Propagating Material and Plants of Fruit Genera and Species 58
   Propagating Material of Ornamental Plants 58
statistics 113, 121, 211
Steinernema 130
   carpocapsae 131
   feltiae 131
sterile insect release technique 132
Streptomyces scabies 175
surveys *see* pest, surveillance
SWARMS 142
Synchytrium endobioticum 17–18, 133, 145, 175, 242

terminology
   EU 51
   phytosanitary 1, 2
test plants 206
Thailand 252
Thenard, M. le Baron Paul 237
Thermus aquaticus 201
Tilletia spp. 165
   indica 222, 224, 226
tissue culture 98, 158–160
tobacco 116
Tobacco mosaic virus 231
Tobacco rattle virus 203
Tomato spotted wilt virus 220
trade routes 4
Trade-Related Aspects of Intellectual Property Rights 40
TRIPS 40

UK 21, 23, 24, 27, 222, 224, 226, 239, 270
   phytosanitary service 85–87
United Nations Environmental Programme 29
University of East Anglia Climatic Research Unit 225
University of Greenwich 141
UPOV 180, 181
Uruguay 254
USA 10, 11, 19, 21, 22, 31, 132, 161, 171, 224, 237, 245, 249, 268, 270, 275
   National Institutes of Health 249
Ustilago spp. 165
   nuda 165

Verdalia cardinalis 130
Verticillium 240

Verticillium – *continued*
   albo-atrum 25, 134
   lecanii 130
Vicia faba 206
Victor Emmanuel III, King of Italy 22
Vinca major 189
Viteus vitifolii 13, 30, 237
   see also *Phylloxera vastatrix*

Wardian case, invention of 5
waste
   composting 242
   disposal 125, 230, 233, 239–244
   liquid 241
   solid 242–244
   management 240–241
weeds 2, 3
World Trade Organization *see* WTO
World War, First 21, 24, 30, 270
World War, Second 23, 25, 48, 270
WTO 29, 39–42, 89
WTO-SPS 29, 32, 38, 40–42, 70, 71, 89, 94, 107, 118, 213, 222

Zonocerus variegatus 137